Around Glare

Around Glare

A New Aircraft Material in Context

Edited by

COEN VERMEEREN
Delft University of Technology,
Faculty of Aerospace Engineering,
Delft, The Netherlands

KLUWER ACADEMIC PUBLISHERS
DORDRECHT / BOSTON / LONDON

A C.I.P. Catalogue record for this book is available from the Library of Congress.

ISBN 1-4020-0778-7

Published by Kluwer Academic Publishers,
P.O. Box 17, 3300 AA Dordrecht, The Netherlands.

Sold and distributed in North, Central and South America
by Kluwer Academic Publishers,
101 Philip Drive, Norwell, MA 02061, U.S.A.

In all other countries, sold and distributed
by Kluwer Academic Publishers,
P.O. Box 322, 3300 AH Dordrecht, The Netherlands.

Cover design: Willemien Veldhoven and Geoff Morris

Acknowledgement of support and disclaimer:

This material is based upon work supported by the European Office of Aurospace Research and Development, Air Force Office of Scientific Research, Air Force Research Laboratory, under Contract No. F61775-01-WF003.
Any opinions, findings and conclusions or recommendations expressed in this material are those of the author(s) and do not necessarily reflect the views of the European Office of Aerospace Research and Development, Air Force Office of Scientific Research, Air Force Research Laboratory.

Printed on acid-free paper

Printed in the Netherlands.

This book is dedicated to the memory of Ad Vlot

Table of Contents

Preface

During September 24-26, 2001, the Faculty of Aerospace Engineering of the Delft University of Technology in the Netherlands organised the *Glare - the New Material for Aircraft Conference*, an international conference on the relationship between design, material choice and application of aircraft materials with respect to new developments in industry. Eminent representatives from the aircraft manufacturing world, including manufacturers, airlines, airports, universities, governments and aviation authorities, were present at this conference to meet and exchange ideas - see the group photo on the next two pages. The fact that the conference was held just two weeks after 'September 11, 2001' put things in a rather unique perspective.

The aim of the conference was to illustrate the many unique applications of the Glare family of fibre metal laminates and to provide for the exchange and distribution of information regarding this material in order to stimulate their acceptance and promote further application.

The introduction of fibre metal laminates into the commercial aviation market took about 20 years' time. Introducing new technologies should not be taken lightly, however; the aircraft industry is by nature rather conservative and innovations must therefore be proven – a paradox actually – in all possible ways before they can be introduced in real aircraft structures. Not only do technical aspects play a role in this respect; historical, cultural, economical and political issues are equally important. So, besides the technical aspects of Glare, which were discussed in the afternoon sessions of the conference and which are published in a first book with the title *Fibre Metal Laminates - an introduction*, the less technical and non-technical issues related to Glare's introduction in aviation were also discussed from different perspectives. These discussions form the contents of this book.

The conference also served as a platform and backdrop to the honouring of Professor Boud Vogelesang, who had long been an enthusiastic driving force behind the development of Glare, as an emeritus of the Chair Aerospace Materials. For this reason, his Emeritus Lecture is also included in this book.

A third book with the title *Glare - history of the development of a new aircraft material*, which provides the inside story of the development activities in the Structures and Materials Laboratory of the Faculty of Aerospace Engineering of the Delft University of Technology, is also available.

Like the previous two books, this book is published by Kluwer academic publishers b.v. Thanks to Arno Schouwenburg once more for transforming our digital copy into a beautiful addition to one's library.

Participants of the Glare - the New Material for Aircraft Conference
in front of the Delft University of Technology Library.

Participants of the Glare - the New Material for Aircraft Conference
in front of the Delft University of Technology Library (continued).

At this point I would also like to thank the sponsors of the conference whose names are included at the end of the book. Without their support it would have been impossible to hold the conference.

Also, I would like to thank the Glare Conference Recommending Committee:

B.A.C. Droste, Chairman of the Netherlands Agency for Aerospace Programmes (NIVR)
Drs. M.C. van der Harst, D.G. for Industry and Services, Ministry of Economic Affairs
Drs.ing. P.F. Hartman, Director of KLM Royal Dutch Airlines
J. van Houwelingen, Chairman of the National Aerospace Laboratory NLR
Drs. A. Kraayeveld, Chairman of FME-CWM
J. Thomas, Senior Vice President Large Aircraft Division, Airbus Industrie
Dr.ir. A.W. Veenman, Chairman of the Board of Directors, Stork N.V.
Dr. N. de Voogd, Chairman of the Board of Directors, Delft University of Technology

I thank you for your confidence in both the material and the conference organisation.

A team of students, who did a marvellous job for the second time, took care of the layout and corrections of the papers:

Ronald van der Meijs
Geoff Morris
Dort Daandels

We all hope that you will quickly graduate.

The cover design is again based on a painting by Willemien Veldhoven. It was transformed into an artistic cover for this book with the help of Geoff Morris.

Finally, I would like to add a few remarks on a tragic event that struck our group. A few months after the conference Ad Vlot, my colleague and co-organiser of the conference, was hospitalised. A terminal disease was diagnosed. Just before the manuscript of this book was ready Ad died on April 18, 2002 at the age of 39. Our sympathy goes out to his wife and three children. We dedicate this book to you Ad, and we thank you for all the work you did during the long period that you were with the Aircraft Materials group.

Dr.ir. Coen Vermeeren
Delft, April 2002

Day 1: DEVELOPMENTS IN AVIATION

Sponsored by: KLM

Keynote lecture

Flying in the New Atlantis - and the evolution of technology

Harry W. Lintsen

Section History of Technology
Delft University of Technology
Eindhoven University of Technology

(Note: The author wishes to thank Frida de Jong and Ad Vlot for their comments)

This article focuses on the long-term trends in technology and takes the book *New Atlantis* by Francis Bacon as its starting point. Bacon lived around 1600 and was a statesman, a philosopher and a scientist. His book was published in 1627, shortly after his death. In this book Bacon presented a society that knew no poverty, was devoid of hunger and free of scarcity and in which people were able to live long and happy lives. New Atlantis was not so much a fairy-tale as a utopia. According to Bacon, such a society could be realised in the future. Some four centuries later we have reached that stage.

Indeed, there are several countries in which these utopian ideals have been realised at this, the beginning of the century. Countries where there is an abundance of food and commodities, where people are protected against the cold and extreme conditions and where, in the centuries since Bacon's days, life expectancy has more than doubled from 35 years to over 70 years. The people in those places are happy, at least compared to those living in countries where this dream is still just a dream. For the first time in the history of mankind whole nations are able to raise their standard of living above the bare minimum.

There are many kinds of utopia, but New Atlantis was the first to put the emphasis on the central role of science and technology, see Figure 1. Bacon's notions about what technology had in store for us were indeed prophetic. New Atlantis used a kind of biotechnology: fattened chickens laid many eggs and the land produced strawberries and other fruits of exceptional size. Food was preserved in cold-storage rooms. Furthermore, people communicated with each other over great distances by means of cables and wires.

Figure 1: Artist's impression of Francis Bacon's New Atlantis.

It is also remarkable that plenty of flying was done in New Atlantis, using machinery that had been crafted in the Machine House. The knowledge underlying this aerospace technology was developed by a scientific community established in Salomon's House, a kind of laboratory. Their work revolved around experimentation, accurate observation and formulation of theories.

Bacon has been called the ambassador of modern science. In his books he formulated the basis of scientific reasoning and the experimental approach. Bacon may also rightly be called the ambassador of belief in technological progress: technology is good for society and so it should have a central position within it.

This simple philosophy has held firm for centuries. The belief in technological progress has never really been disputed. Unlike feudal structures, religious disputes, class differences and capitalistic attitudes, technology usually remained untouched. In the twentieth century, however, this position radically changed with modern technology and modern society coming in for heavy criticism. In the seventies, a definite end to the belief in progress was heralded. Nowadays, the remaining 'believers' are looked upon as being naive and irresponsible.

All of this leads to two remarkable paradoxes, which are:

1. As soon as science and technology fulfil their promises, people lose faith in them, or to put it another way: countries that realise the New Atlantis unleash mass criticism of modern technology. Even the aerospace industries were not spared. An example is the fiasco surrounding the race for supersonic transport, i.e. the American SST project and the European Concorde.
2. The more we succeed in controlling nature and society by technology, the more vulnerable society becomes to human behaviour. The terrorist attacks in the United States two weeks ago (ed.: September 11, 2001) are a horrible example of this paradox. I will return to that subject later. In general, New Atlantis proves to be a risk society with several kinds of risks through complex technological systems within communication, the energy industry and also aerospace.

I would like to examine these paradoxes more closely. First, I would like to establish when it was that we were first able to pass through the gate and enter New Atlantis. What were the technologies and intentions of that period? In the second place, we must ask: 'Why did the creation of New Atlantis mark the end of a belief in progress?' The chief question, however, is: 'How must we now believe in technology, especially in the aerospace industry?' In other words: 'What should be our intentions with regard to (aerospace) technology as we glide further into the twenty-first century?'

The technological revolution

Technological expansion has been explosive in recent centuries, especially after Francis Bacon's time. Before then, however, accelerations in technological development could also be detected, see Figure 2.

The first acceleration was the technical development that took place during the time of the Agricultural Revolution. The First Agricultural Revolution took place between roughly 8000 BC and 6000 BC and the Second Agricultural Revolution between 5000 BC and 3000 BC.

The Second Agricultural Revolution was concentrated in the Near East (in Mesopotamia and Egypt) and in the Far East (in India), see Figure 3. The main innovations were large irrigation systems, metal processing and the construction of temples, palaces and bridges.

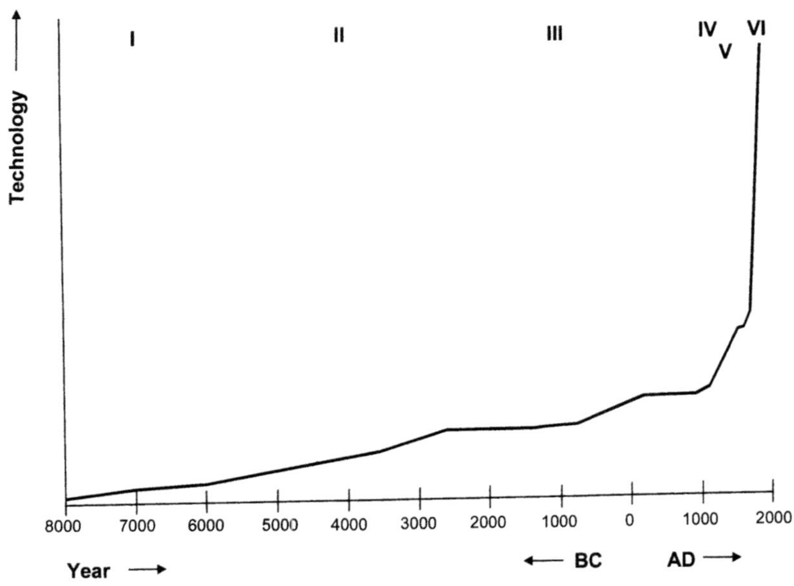

Figure 2: Accelerations in technological development through the ages.

Figure 3: Construction of temples in Egypt during the Second Agricultural Revolution.

A third acceleration in technical development took place during the time of the Greeks and Romans (during the period of Classical Antiquity between 600 BC and 400 AD). Progress was chiefly made in the fields of philosophy, the natural sciences and in law and organisational areas. However, great achievements worthy of note were also being made in the field of technology, notably in the areas of shipbuilding, navigation, infrastructure and military technology, see Figure 4. Flight was still not possible at this time, although the dream of flight was already strongly present. All the civilisations of the time show many signs of man's flight on both mythological and physical wings, e.g. the Greeks had Daedalus, the Peruvian Indians had Ayar Katsi – the flying man, the flying carpet was a popular image in the Arabian world and the Christians saw angels. It is also worth noting the fact that not all the images of flying people were positive. Flight could also be the symbol of mankind's folly such as in the Greek myth of Icarus, who used his father's invention to try to reach the sun - with horrible consequences. The main reason for his failure was that the glue used to manufacture the wings was not sufficiently heat-resistant. Glare, be warned!

There was a remarkable technical pace of change during the Middle Ages, the period known in Western Europe as the Dark Ages. China rather than Western Europe then stood at the forefront of technological progress with innovations such as the printing press, the compass and gunpowder. Eventually, Western Europe did take on board some of the important discoveries during the later Middle Ages. It is suspected that the first attempt of human flight was made during this period, see Figure 5. An English monk attached wings to his arms and legs and flew an alleged two hundred metres, although he broke both legs and was paralysed for life after a hard landing. In the Middle Ages, China was pursuing a different – and ultimately better – method, namely the use of kites (a kind of fixed wings) and propellers.

Figure 4: Roman aqueduct.

Figure 5: Artist's impressions of the first attempt at human flight by an English monk and of China's pursuit of using kites and propellers respectively.

During the Renaissance, the time when Francis Bacon lived, technological development stagnated. This period of widespread creativity was not so much linked to technology as to a revolution in art, the emergence of a new human and world view and the advent of modern science. In terms of technology it is important to see that innovation was positively valued by the culture, and therefore the role of the inventor and the engineer became central.

After 1750 technology developed exponentially during what was, as everyone knows, the era of the Industrial Revolution. The Industrial Revolution took place in phases. We refer therefore to the First, Second and Third Industrial Revolutions.

The First Industrial Revolution took place in England between 1770 and 1830, and steam, the textile industry, iron and the railways were its central technological components.

During the Second Industrial Revolution, which occurred between 1870 and 1914 in the USA and Germany, the electronics and chemical industries became the main catalysts of technological change.

The Third Industrial Revolution is now fully underway, with the USA in the lead and information technology (IT), new materials and biotechnology forming the spearheads of progress.

The First Industrial Revolution is noteworthy for us, since it marks the opening of the doors to the New Atlantis. Steam power enabled modern man to dramatically increase productivity and the availability of food, thereby raising the minimum standard of living considerably. As a result, the life expectancy of the average Dutchman saw steady growth from 1860 onwards, see Figure 6. At that time, life expectancy was 35 years but has more than doubled since.

Aerospace played no role in these changes. However, some improvements in aerostatics, aerodynamics and aeromechanics did occur during the First Industrial Revolution. The result of these improvements was embodied by the first flight of a hot-air balloon by the French Montgolfier brothers in 1783, see Figure 7.

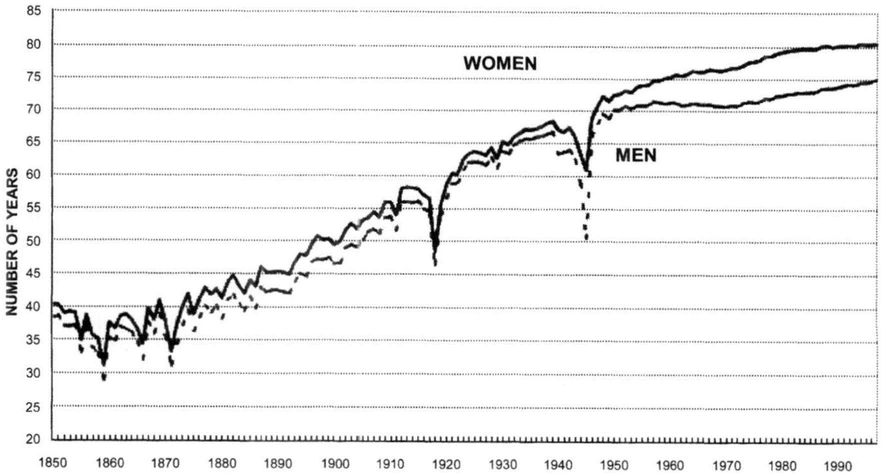

Figure 6: Average life expectancy in the Netherlands 1850-1997.

The Second Industrial Revolution: The birth of New Atlantis

The most important characteristic of the Second Industrial Revolution was that it involved the strategic application of science in production processes. Bacon's dream of a close relationship between science and technology was realised when the German chemical industry emerged and when the first laboratories were set up and systematic research was carried out. Companies in the USA followed this example and the modern chemical industry was born.

Exponents of the electronic industry were also among the first to make use of laboratories. They applied their scientific knowledge of electricity and magnetism to electric lighting, telegraphy, telephony and power.

Figure 7: Artist's impression of hot-air balloon around the time of Montgolfier.

Together with the work of individual inventors, scientists and engineers this research led to a series of scientific and technological breakthroughs around 1900. These included the fields of physics (e.g. the discovery of X-rays in 1895), medical science (among other things chemotherapy, brain surgery and vitamin systems), communication (telephony, film and wireless telegraphy), the energy industry (especially the electrical power industry) and transport (aircraft, bicycles, cars, trams and local rail networks).

The first aircraft to fly was not developed in the laboratory. Instead, it was the product of a group of individual scientists and engineers and the modern knowledge network that linked them together. The development of the modern aircraft industry began with the work of the English researcher George Cayley (1773-1853) during the beginning of the 19th century and ended with the first powered flight by the Wright brothers in 1903. Many different people contributed to this process of development during this time. They carried out theoretical research, conducted experiments in wind tunnels, developed prototypes, designed and flew gliders, etc. The results of their work were published in magazines and books and thus a large database of knowledge was built up and made accessible. Experiences were exchanged at meetings held by scientific societies, such as the French *Société d'Aviation* (1863) and the English Aeronautical Society (1866). In other words, a professional community had sprouted up and this community was busy making mankind's dream a reality, see Figure 8.

Figure 8: Early days of aviation; the English Aeronautical Society.

The social debate

In short, the *fin de siècle* of the nineteenth century was an exciting period and that was how people living at that time experienced it too. Mankind was on the brink of a great revolution. People counted the blessings of technology and felt that nature could definitely be conquered and people could be freed from their 'vale of tears'. Similar predictions for the twentieth century were forecast by various social groups (Catholics, Protestants, liberals, socialists and anarchists), by entrepreneurs and workers, but also by economists, lawyers, artists and writers. Society was in debate about the future.

One professional group that was also completely involved in the social debate was the engineers. They presented themselves as the guardians of modern technology and claimed, on that basis, a certain social status. They were able to harness the forces of nature and make such forces useful to mankind. They were also equipped to deal with the chaos created by the First Industrial Revolution and by capitalism.

They lobbied for reform, intervention by the government, social legislation and scientific management. They developed technical facilities for public housing, public health and general hygiene, e.g. see Figure 9. Engineers became heavily involved in the social debate, so much that one could even speak of there being an engineers' revolution. One century later one may conclude that the engineers did fulfil their promises. Anyone born around 1900 can confirm this.

Take, for example, my grandfather and consider what he saw in his lifetime, e.g. the first cars in the Netherlands, the first aircraft, radios and televisions, the first electric irons and ovens, the first showers with warm running water, the first supermarkets and fast food, etc. My grandfather lived to be 87 and for most of his life he enjoyed good health. In the Netherlands he was one of the first to pass through the gates of New Atlantis. Indeed, that is how he must have experienced it, because in 1900, at the age of four, he was taken to the World Exhibition in Paris, see Figure 10.

Figure 9: Engineers developed technical facilities for general hygiene, among other things, around 1900.

Figure 10: My grandfather must have felt like one of the first to pass through the gates of New Atlantis, when at the age of four he visited the World Exhibition in Paris in 1900.

The inferno of New Atlantis

The experience of going to the World Exhibition must have been quite overwhelming. Forty countries took part in it. There were 83000 entries, 48 million visitors from all over the world and some 150 international congresses had been organised. The blessings of modern technology were not disputed. Technology brought progress.

There was, however, another side to the World Exhibition. Eight columns in the manifestation's catalogue were devoted to listing innovations pertaining to the field of military technology, i.e. new types of cannons, new explosives, new warships and new tanks, etc. Aerospace was also represented by the presence of military balloons. These were mostly intended for viewing and observing the enemy and for targeting artillery fire and bombardments. The aeroplane was also quickly applied for these tasks after the Wright brothers had made their maiden voyage. From the very beginning the aircraft has been strongly linked with the military system and pilots were looked upon as the new heroes of war, see Figure 11.

Around 1900 there were already visible signs that an arms race was developing. The world was rapidly changing. The Hapsburg Empire was crumbling, as was the Ottoman Empire. The Balkans had become a political battlefield and Germany had become an impressive power. In Asia, Japan was trying to secure power and was at war with China and Russia. Increasingly it was rising nationalism that was dominating the world stage. The prelude to two world wars was underway. The twentieth century will surely be remembered as the bloodiest century ever, in which modern technology played a crucial part. This was a development that no one had predicted.

Furthermore, the twentieth century has been a century that has left working man deeply scarred. Man has always had to adjust to changes in the working process. Production techniques have been renewed and industries have undergone reorganisation. It was a century in which functions changed and knowledge and skills became outdated. People often lost their jobs. The economic depression of the thirties led to a trauma that is still recalled in times of mass unemployment.

The twentieth century was also the century in which man was once again confronted by nature, though in a completely different way this time. Two major oil crises, dead rivers, stifling smog, exhausted soil and dying landscapes exposed nature's and therefore also man's and society's vulnerability. The twentieth century has been one of contrasts but technology has always remained at the forefront.

Figure 11: From the beginning, aircraft have been linked with the military system and pilots were looked upon as new heroes of war.

What then can history teach us for the future?

We started this lecture by going back some four hundred years, to the utopian Francis Bacon and his New Atlantis. His utopia has been realised but Bacon would be disappointed with the results because the twentieth century has been too much one of contrasts. What can we learn from this? I can define six lessons.

Lesson 1:

The twentieth century was a technological century. Science and technology have become, as Bacon forecast, one of the main vehicles of social change. Scientific and technological developments have influenced social and global developments in profound ways.

Lesson 2:

History provides us with three views on socio-technological development, each of which has, up to a point, general validity.

The first view is that technological progress is autonomous. Science and technology stimulate each other in a stream of perpetual innovations. Key technologies pan out, form large-scale systems with their own dynamics and penetrate all corners of society.

Air transport in the twentieth century is a good example of this. Born from the interaction of many different scientific and technological disciplines, pushed forward by an enthusiastic public, strongly linked to military and economic interests and looked upon as a matter of national pride. As an inevitable result aircraft became larger and faster and every 'self-respecting country' developed its own large-scale air transport system. In many ways this was also a false result, i.e. aircraft industries, airlines and airports were often supported by huge subsidies, building contracts and other government support.

The second view holds that technology involves making choices. Technological development presents us with a lot of possibilities from which to choose.

Once again, the aerospace industry supplies many examples of this. Take the debate about Amsterdam's Schiphol Airport – one of the most important social debates running in the Netherlands at this time (ed.: September 2001). In 1995 the decision was made to construct a new fifth runway. Yet the air transport industry is so dynamic that there is already talk about further expansion. Many alternatives present themselves, e.g. a sixth and seventh runway at Amsterdam Airport or the construction of a new airport in the sea. Another alternative is to limit the growth and to stagnate the expansion of air transport in the Netherlands. Each of these choices will have consequences for the economy and the quality of life for the citizens.

The third view is that developments in science and technology are unpredictable, as is the interaction between technology and society and, indeed, the social outcome of such interaction.

Around 1900 no one would have predicted that a world war making use of hot-air balloons and around ten thousand aeroplanes would take place just around the corner, resulting in the death of around fifty thousand pilots and other associated personnel.

Around the year 2000 nobody could have predicted a terrorist attack such as that which took place in America on September 11, 2001 with more than five thousand victims and far-reaching consequences for aerospace, society and international relations.

Another example is the V/STOL (Vertical/Short Take-Off and Landing) aircraft, which was predicted to have a glorious future in the 1950s and 60s. These aircraft were able to take off and land vertically and would therefore be able to fly right into city centres using only small airports, such as the roofs of railway stations and car parks. The manufacturers had not anticipated the ensuing public protest with regard to noise, pollution and the risk of accidents. Partly as a result of this, the concept failed on the whole despite the construction of more than fifty prototypes. Visions of the future rarely see reality in the aerospace industry due to the unforeseen and unintended side effects they generate.

Each of these three views contains some truth, which would indicate that socio-technological development is a complex process. It places demands on our attitude towards technology, which brings me to the third lesson that history can teach us.

Lesson 3:

Be utopian, but accept complexity and remain open and flexible. Bacon's utopia was too simple. For him science and technology stood for progress by definition, and this is still the dream of many engineers who hope that technological progress will automatically lead to a better society. But this is an illusion. Scientific and technological developments are too complex for that, as indeed are human beings and social, political and economic processes. Accept uncertainty and learn to anticipate unexpected developments. This means working to create a society that is open and flexible. It also implies that the technological systems that are designed and constructed have to be flexible as well.

One project in which this failed completely was that of Concorde. Many people firmly believed in the need for faster transport over longer distances during the 1950s, which led them to the conclusion that a supersonic aircraft was required. The development and construction of the Concorde between 1960 and 1975 required 21000 workers and 500 suppliers. During this development period, the market segment for which the Concorde was intended changed radically. Before the first

Concorde could even take to the skies, the project was already out of date but could no longer be stopped. The Concorde may have been a technical leap forward, but remains an organisational fiasco, a financial disaster and a social débâcle.

Lesson 4:

Be optimistic, but think pessimistically. We need utopias, in the sense of ideals and future projections to motivate us and steer our ambitions. At present there are challenges enough. In the greater part of the world New Atlantis has still not been realised. War, terrorism and violence constitute a threat to the world as a whole and to the stability of individual countries. A stable society in which people can live in harmony with nature for generations has yet to be created, but let us remain realistic.

Take the Airbus A380, for example, the new pride of the Airbus fleet that will depend largely on the application of Glare and other new materials. Airbus calls this 'super swan' the Green Giant. It will – it is promised – be quieter (than the Boeing 747), be more fuel-efficient and produce fewer harmful emissions. In general, the last decades have brought noise, energy and the environment to the forefront of aircraft development. It would be rash, however, to assume that this means that the aerospace industry automatically contributes to a sustainable society. Many of the improvements in sustainability are being nullified by the growth in this sector.

My motto would be: 'Think pessimistically and formulate boundaries – permanent or otherwise – for technological development.'

Lesson 5:

Although the pace of technological development is fast, take time to consider changes. This might sound contradictory. We live in hectic times; competition is sharp and it is important to react quickly in order to survive. Still, despite this, I would recommend taking time to contemplate change. Research has shown that the decision-making processes for large-scale technological projects in countries such as England, Germany and the Netherlands take some 15 years on average. I would say 'rightly so!' Take time to listen to others and learn from others. Take time to list the various interests, to develop alternative plans and to experiment with unexpected solutions. Be holistic and integrate all the various values into your new designs. It will demand enormous effort to correctly channel technological developments, certainly in a period of socio-technological revolution like that of the present.

Taking time to make decisions does not guarantee their validity, however. Much depends on the quality of the decision-making process. Personally, I find it a disgrace that the Dutch populace was unable to find a satisfactory solution to the problems surrounding Schiphol, an airport that ranks among the world's top in terms of technology and facilities. After thirty years of debate we are still using an airport set in one of the most densely populated regions on Earth. Unfortunately, we can see nothing but missed opportunities, short-term politics and a lack of nerve and vision when looking back on this matter.

Lesson 6:

There is a new fundamental dilemma that has arisen for modern man. The question is not only: 'Can we do what we want?' but also: 'Do we want to do what we can?' Francis Bacon and the generations that came after him knew what they wanted, i.e. a New Atlantis in which the problem of poverty was resolved. Their question was: 'Do we have the technical possibilities to achieve that?' The technological possibilities do exist now through mass production, mass consumption, large-scale systems and economic growth. The question for modern man is: 'What next?' The technical possibilities are endless.

Both questions have come very much alive at the moment with regard to the terrorist attacks in the United States on September 11, 2001. We want a democratic and prosperous world, but the question remains: 'Can we do what we want?' I do not think the answer can be found in technology, like in Bacon's time, but in human relations and values such as equality and tolerance. Furthermore, the reactions to the terrorist attacks can be very different. The United States and the Western world have several technological means at their disposal to react with. Do we want to do what we can? I hope that the answer to the recent terrorist attacks will not be a third world war or the use of nuclear weapons, but a war against terrorism.

In general the rule applies that in the 21st century new questions will be asked. We can create many worlds, but the question will be: 'What kind of world do we want?'

New Atlantis is something that has to be rediscovered. That is our challenge when facing the coming century.

Response 1

Sustainable aviation: KLM's view on 'Flying in the New Atlantis'

Udeke N.J. Huiskamp

KLM Government & Industry Affairs
KLM Royal Dutch Airlines

Professor Lintsen addresses in his lecture a key question in aerospace technology, i.e.: 'What should we achieve in the 21st century?' As a user of technology, KLM keeps a close eye on technological developments and is willing to contribute, within the bounds of its ability, towards sustainable aerospace technologies.

What has been achieved?

Air transport has become the most successful mode of public transport supporting global, regional and local economies. Aviation is crucial for trade and tourism and is generally felt to be essential to people's quality of life. This success is based on technology in combination with beneficial macro-economic and sociological conditions. The aviation industry has grown by around 9% per year since the 1960s and is expected to continue growing, making it one of the fastest-growing global industries. As a result of technological and operational improvements, tremendous progresses have been made in the areas of fuel efficiency and safety. Aviation is the safest mode of transport, as the absolute number of fatalities per passenger-kilometre is much lower than with alternative means of transport. As a result of the focus on cost reduction and environmental performance the efficient use of kerosene has been a guiding principle in both the development of aviation technology and in day-to-day aircraft deployment. As a result modern fleets are 65% more efficient than in 1970. On average, KLM uses 3.5 litres of kerosene to transport a passenger over 100 km! Of course fuel efficiency decreases on short-haul routes[1], but the energy consumption per passenger-kilometre of an aircraft is still comparable to that of a modern car over such distances.

[1] On the Amsterdam-Paris route KLM uses approximately 6 litres of kerosene per 100 km to transport 100 kg. A passenger with luggage weighs ±100 kg.

It is expected that a further 10% improvement in fuel efficiency will be realised over the next 10 years, see Figure 1. However, these improvements do not fully compensate the growth in air traffic. The net effect is an increase in pressure on the environment through aviation. The contribution of aviation to the emission of anthropogenic greenhouse gases is expected to increase from 3.5% to 5-6% in 2050.

There is no full understanding and consensus among scientists about the contribution of all aviation emissions to the greenhouse effect. A huge amount of work needs to be done to gain a better understanding of the effects of aircraft emissions on the global atmosphere. Despite the uncertainties about the precise effects, the precautionary principle requires that the international society should take all reasonable measures to stabilise greenhouse gas concentrations. This brings us closer to answering Professor Lintsen's question: 'What should aerospace technology achieve in the 21st century?'

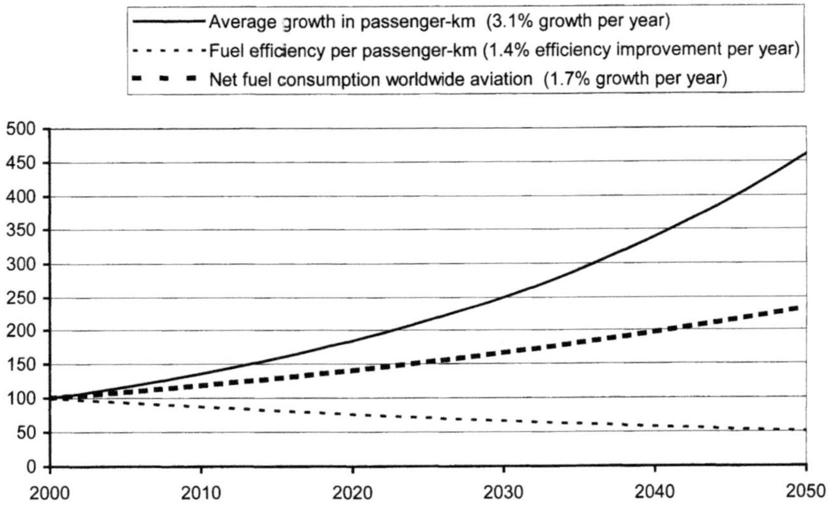

Figure 1: Forecasts of worldwide aviation demand and fuel consumption (index: 2000 = 100) [source: IPCC-report 1999].

Now that the standard of living in many countries has achieved the status of the 'New Atlantis', we realise that our way of life might compromise the ability of future generations to meet their own needs. At the 1992 United Nations Conference on Environment and Development in Rio de Janeiro the international community acknowledged this by adopting the 'Agenda 21' (Agenda for the 21st Century) with a focus on sustainable development.

For KLM, sustainability involves the simultaneous pursuit of economic prosperity, environmental quality and social equity. KLM is presently exploring what sustainability implies in terms of day-to-day operations, which objectives will be given priority and in what time-frame these objectives may be achieved. Although this quest has just begun, it is clear that sustainability requires a more proactive attitude and the exploitation of environmental-commercial win-win opportunities. New materials, such as Glare, present such opportunities. Other areas of development make a less obvious contribution towards sustainable aviation. New aircraft types, such as the Airbus A380, may be more fuel-efficient due to scale effects and may save fuel through a reduction of congestion at airports. On the other hand, they may also cause overall flight distance to increase, as more passengers have to transfer at hub airports since it is not feasible to fly directly (and consequently over less distance) to the desired destination.

It might be possible to accommodate aviation demand using smaller aircraft, like Boeing's Sonic Cruiser. However, it is not clear to which extent sustainable aviation might benefit from this type of aircraft, which is designed to cruise at higher altitudes with higher speed. The residence times of some aircraft emissions (NO_x) are known to increase with altitude and there are indications that high-altitude flight contributes relatively strongly to the greenhouse effect.

KLM is well aware that, as a market participant, it is capable of taking into account environmental considerations when buying materials and products. Of course, KLM also takes into account environmental factors when acquiring a new aircraft. In doing so, we indicate to our suppliers that we want them to develop environmentally sound products. Apart from being well-informed on the environmental performance of products, KLM aims to intensify the debate on sustainable technologies. Although the actual influence that KLM can exert differs from supplier to supplier, progress has been made in the use of less hazardous substances in aircraft maintenance, for instance.

KLM's proactive sustainability strategy ensures that we view new concepts and technologies in the aviation industry from various angles. Sustainable aviation requires a holistic view, in which the ever-important emphasis on safety and cost-effectiveness are combined with environmental performance. This can only be realised by a close co-operation between scientists, technologists, industry bodies and airlines. The role of aviation in modern society and the need for sustainable development are both too important to let them proceed at random. They must be driven by choice instead.

Sustainable aviation is not a utopian dream, but can be achieved by making choices, by implementing effective measures, by an open exchange of information, by focusing on long-term objectives, and last but not least, through the implementation of new technology.

Response 2

Ben A.C. Droste (Lt Gen ret RNLAF)

Netherlands Agency for Aerospace Programmes

I thank Professor Lintsen for his highly informative lecture. I am personally an avid student of history, but in my roughly 40 years spent in the aerospace world I notice that people such as Professor Lintsen are a rare find. I think I also have an explanation for this phenomenon. People that choose careers in aerospace share characteristics that are not favourable to looking backwards. Whether an engineer or a pilot they tend to love technology and are great believers in the dictum that everything can be bettered by introducing new technologies. They are not often of a romantic and introspective nature. I think I can prove my point by stating that in our ranks we do not count many artists such as painters, writers or composers. Of course there are exceptions and I personally am a great fan of the writings of the famous Saint-Exupéry. However, I have to admit that as far as my knowledge goes he was not a great flyer. He made many avoidable accidents, even when the state of reliability of the aeroplanes in his days is taken into account. As an example I cite one of his many tries to set world-records. In a race from Paris to Saigon he ran out of fuel and had to crash-land in the Egyptian desert because he had badly prepared himself on the available weather data. Only through a miracle was he finally rescued. While this made an excellent story it also underlines my point that people in aeronautics have to be less romantic and more rationale based.

This does not, however, mean that we in aeronautics have to neglect or discard history. Professor Lintsen did a good job with that. Of course you might remark that at least we take recent history at heart. Most certainly a lot of lessons will be learned and will be applied following the most horrific drama, in which four passenger aeroplanes were involved in the American skies (ed.: terrorist attacks on September 11, 2001). These lessons will lead to new rules and regulations and new innovative technology will be applied to prevent a repeat of the disaster. Already we see glimpses of what that can be, e.g. flight programmes that make it impossible for an aeroplane to hit anything other than a planned runway. These are ideas that have already been proposed by Professor Mulder of the aerospace faculty here in Delft, ideas that could already make use of available technology, as we can see in military aeroplanes that fly blind at night at 250 feet in their 'terrain following' mode. These corrective actions are all well and necessary, but I side with Professor Lintsen in the need to take a look over a longer period of time. Too often we see that accident investigation leads to a new set of rules and regulations that address the past accident instead of anticipating what is to come. I compare this with an example from ordinary military life. As you might recall from the days you spent in military service, it is forbidden to put your hands in your pockets when addressing each other. When this rule is breached, investigators and committees often come up with ideas like: 'Let us do away with pockets', while it is far more effective to address situations like these by analysing and addressing human behaviour.

What can we see in aircraft development in the longer term? Will technology move forward with the incredible speed we have seen in the last century? Will it once again be possible to make quantum leaps from the Wright Flyer of 1903 to the fly-by-wire controlled F-16s and civilian aeroplanes of today? What do we glimpse of the future? I follow the lead of Professor Lintsen by being optimistic but thinking pessimistically. I am optimistic in hoping that the joy of piloting an aeroplane will remain, as I have experienced for thirty-eight years in the Royal Netherlands Air Force. Flying a fighter like the F-16, as technologically advanced as it is, is still a challenge to the individual skills of the pilot. In a dogfight between two fighters the winner is a direct result of the quality of the pilots concerned and even today a very good fighter pilot stands a good chance to beat opposition equipped with a better system. With his situational awareness he can still beat most of the opposing weapon systems. Seeing a threat such as a missile coming in still gives you a fair chance to outmanoeuvre it. However, thinking pessimistically as advised by Professor Lintsen I am aware that this human difference is rapidly eroding. The newest fighter designs are not so much aeroplanes that can make even tighter turns than the F-16 with its 9 G., on the contrary, the design criteria now focus on the effectiveness of the whole system. That means the effectiveness of sensors to recognise and identify threats at such an early stage of the engagement that you can fire your very smart missile while the situational awareness of the opposition is still at a loss. When you have to make defensive escape turns a lot of things have gone wrong in this new generation. Systems will more and more take over the role of the aviator, as much as I hate it from an emotional point of view.

While the next-generation fighters will still make use of the fighter pilot in a high-tech cockpit, in new-generation airliners this is not in fact necessary any more. Auto take-off, flight and landing systems will take care of the complete flight, as is already the case for 95% of the flight time in the present-generation airliners. Those systems will perform far more effectively than a human pilot will. You might know that auto land systems are already designed, if not always certified, to land at higher crosswind speeds than the pilot is allowed to do. What do we see in practice? Most pilots and their aeroplanes are prohibited to land in crosswinds above 25 knots. So under higher speeds aeroplanes either have to divert or can not take off, costing a lot of money for their companies. However, do we see pilots using their auto land systems at 25 knots? Generally we do not, in particular when the wind is gusty. The reason is that pilots instinctively think they can do it better than the system. Is that really true? I would argue this is not the case. How often does a normal pilot fly in severe gusty crosswinds? There is a good chance that he has not experienced it for some time and will have lost his seat-of-the-pants feeling, as pilots like to call it. So he is trying to compete with a very capable auto-land system that does not need recent experience. Whatever experience there is, is included in the software that has been designed and upgraded all the time by qualified engineers like many of you in this room. So the sensible thing to do is to trust the system and let the pilot be a systems-checker and I am sure this will happen sooner than pilots would like. I could cite similar examples for air-traffic controllers who notwithstanding their modern computers still only trust themselves, thereby insufficiently utilising the possibilities of the system. How long will airlines accept that on a bad, misty winter day the company's production is reduced by 50 or 80 percent? Well I say that all of them do, because they do not yet make full use of automated systems. Can you think of a company that has to accept such losses in productivity inadvertently and where the company leadership apparently accepts these facts? Well I cannot, but this is the case in the aviation industry, civilian and military to be clear.

Does this mean that I am a convinced believer in technological answers to all challenges? Again I side with Professor Lintsen. It is good to think pessimistically in this respect. It will provide better solutions in the end. However, never be too conservative in accepting new solutions, as many pilots and air-traffic controllers tend to be.

Finally, I proudly cite the example of Glare, i.e. the new technology in aircraft materials that has brought us together here. It is thanks to the persistence of the very few that were here in Delft 22 years ago that it all started and that we now find ourselves at the dawn of a new revolution in aircraft materials. How else could you describe the decision that now, for the first time since 1932 when we changed from fabric to aluminium to cover aeroplanes, we are at the beginning of large-scale applications of Glare in the newest-generation passenger aeroplanes. So these few inventive engineers were not conservative at all. I conclude by paying my respect to these great engineers and all those who believed in them. You have done a great job!

Response 3

Pleading for a vision

Heinz G. Klug

Airbus Deutschland GmbH

(Disclaimer: The author wishes to state that this paper represents his personal views, which are not necessarily identical with the position of Airbus)

Harry Lintsen has put the development of aviation into the great historical perspective. He has pointed out where we succeeded during the last decades, and where we failed. He has discussed the complexity of decision making and the difficulty to choose the right target.

My contribution will not be so well balanced. We must learn from the errors and mistakes of the past, yes, but all our activity can only be aiming at the future. So I pick up Harry Lintsen's advice, i.e.: 'The New Atlantis must be reinvented.'

My contribution is a plea for a vision and a plea for a certain way to approach that vision. It is the vision of somebody who has worked in aviation with enthusiasm all his professional life, has seen aviation thrive, and is convinced that aviation can have a grandiose future. The *vision* – or the *mission* for the industry, if you prefer – is:

To achieve long-term continuing growth of civil aviation until every man and woman on earth can fly as often and as far as they want and, when doing so, do not harm other human beings, or the environment.

Civil aviation has enjoyed continuing growth over many decades at rates over 5%, albeit with a lot of short-term hiccups, which often caused hectic reactions on the side of industry and even panic. The big players in the field predict further growth at rates between 4% and 5% per year. A wonderful perspective for the aircraft industry; our business will grow for a long time to come. Will it, however?

Actually, growth itself is our business. This is easy to demonstrate using a simple 'model'. Let us consider a 25-year period, during which air traffic is growing 4.5%

per year. That means that traffic will exactly triplicate in the period and hence the number of aircraft – or rather seats – to serve the traffic must triplicate. If there were 100% aircraft at the beginning of the period, there must be 300% at the end. 25 years is the average service life of an individual aircraft (tendency: increasing). So, the 100% aircraft in service at the beginning are all replaced just once in the 25-year period. 300% will be in service at the end and obviously 200% has been produced to serve the growth. If some event would stop growth, we would lose two thirds of our business. We would be back at the production rates of 25 years ago – a catastrophe indeed. So the industry has a really essential interest in continuing growth, approaching some asymptote in the long run, of course. However, is continuing growth a realistic possibility?

It can be demonstrated that the level of air traffic in some country, e.g. expressed by the number of flights per capita and year, is proportional to the economic wealth, e.g. measured by the gross national product per capita – although there are more factors of course, such as geography. So, if we foresee long-term world-wide economic growth, we can hope for continuing growth of air traffic. Today less than 5% of the world population, i.e. the USA, produces some 40% of the world's air traffic. Approximately 40% of the world population, i.e. India and China, produce less than 4% of the world's air traffic. What a fantastic potential! A rough estimate says that air traffic would have to grow by a factor well above 10, if sometime in the future everybody in the world were to fly as much as US citizens do today. However, is that a realistic scenario?

A closer look shows that there are many drivers, but also many potential obstacles to such growth. Increasing wealth and increasing population are probably the strongest drivers, but here the first questions arise. Can we have both at the same time? Is population-control not a prerequisite to achieve 'Western' levels of economic wealth? Globalisation of economy is at least partially a product of cheap flying, but is also a strong driver. The spreading of our Western lifestyle, our quest for pleasure, the attraction of exotic countries – which is a matter of perspective of course. You can name many drivers, some of which may be built into the very nature of man.

It may not seem probable, but we can also imagine a contrary trend, a spiritual renewal or a world-wide change of basic values in people's life, which makes travelling less attractive. We can speculate that the new communication technologies will save a lot of trips – I doubt that this will be the net effect. We can envisage that cyber-worlds delivered per Internet and artificial paradises under giant glass domes will replace vacations in the real world, cutting down the market for pleasure trips. We can not exclude a new political fragmentation of the world – we may even see the beginning these days – political catastrophes, economic stagnation or even a complete breakdown of civilisation as the Club of Rome predicted. However, let us not wait for catastrophes to come. That sort of anticipation will paralyse us. It is better to take an optimistic view of the future and work hard to make it happen. Sometimes limited resources (fuels, materials) or air-traffic system saturation are quoted as putting an end to growth, but I can not see that either. There is a lot we engineers can do.

The strongest potential obstacle to the 'unlimited' growth of civil aviation, a 'danger' which now is right around the corner, is the need to protect the environment. Sometimes I feel that we as an industry react to this 'danger' like the well-known three Chinese monkeys. Sometimes the perception of the problem appears to be a generation problem.

It is certainly true that aviation today contributes only some 3% to the anthropogenic greenhouse effect. But that does not mean we can ignore our emissions. We can not simply disclaim responsibility for our share of the global problem. As the special IPCC-report *Aviation and the Global Atmosphere* is pointing out, aviation contributes to the greenhouse effect in several different ways, and most are not very well understood and can not be quantified with a high accuracy today. But of all emissions, only carbon dioxide (CO_2) has a true long-term effect. If you would stop all aviation today, the effect of nitrogen oxides (NO_x), of contrails or of changes in general cloudiness would be gone within a few days or weeks. But the carbon dioxide we emit remains for a typical residence time of 100 years. What we emit today we leave for our great-grandchildren. That is why we must worry about carbon dioxide more than any other emission. What control do we have over carbon dioxide emissions?

The amount of carbon dioxide emitted is strictly proportional to the amount of kerosene burned. Now we are all clever engineers. We continuously improve our engines and our aircraft. What are the prospects? My personal view is that there is little to come in aerodynamics other than laminar flow wings, which is a doubtful proposition due to the complexity in structure and systems required. I have many doubts about unconventional configurations like 'blended wing/body', a concept now *en vogue* (once again). On the other hand, I do expect significant progress in structures and materials, e.g. widespread introduction of clever composites such as Glare. But alas, possible improvement will not be enough. We have improved fuel efficiency of our aircraft by approximately 2% per year during the last decades (two thirds thanks to improved engine technology and one third by airframe technology). To maintain that rate of progress over the next decades will be very difficult, if not outright impossible.

Now, if we want air traffic to continue growing at, say, 4.5% per year and improve fuel efficiency by 2% per year, carbon dioxide emissions will grow by 2.5% per year. How long can we get away with this?

The need to reduce the emission of carbon dioxide (and other gases) is now generally accepted (Kyoto, Bonn), even though the mechanism to achieve – to enforce! – such reductions is not yet well defined and even though the biggest player, i.e. the USA, is still standing aside. It would be foolish to hope that the problem will go away just like that. Most people will also agree that it would be foolish to assume that aviation will get a complete exemption from the need not only to stabilise, but also to reduce carbon dioxide emissions. Of course we can speculate that we will always be able to buy the right to emit, e.g. by paying ever-increasing taxes or by means of emission trading.

However, is this a good strategy? As an engineer – and as an ordinary citizen – I find that this approach is not satisfactory. We should look for technical ways and means not only to reduce, but also to completely avoid the emission of carbon dioxide. That is the proper vision and mission for an engineer!

Effectively avoiding carbon dioxide emissions could be achieved by using kerosene based upon biomass. Technically speaking this should be possible. Nevertheless, I have my doubts about bio-fuels. Will the production of bio-fuels (on a really large scale) not compete with production of food? Will the change to bio-fuels not re-create the current situation that a few countries dominate the market, because they alone have the resources?

Another recipe could be to filter the trace gas carbon dioxide emitted by aircraft out of the atmosphere again, split it up into carbon and oxygen and combine the carbon with hydrogen produced on the basis of renewable energies to form a synthetic kerosene. Feasible in principle, yes, but probably very expensive!

The most promising new energy carrier in my view – many, but not all, will agree – is hydrogen, i.e. liquid hydrogen (LH_2), when it comes to aircraft.

Liquid hydrogen offers great advantages:

- It can be produced on the basis of any renewable energy anywhere in the world through electrolysis of water, but it can also be produced by gasification of biomass.
- When burned, the only primary product is water again, i.e. a closed cycle.
- It contains nearly three times as much energy per weight unit than kerosene, a fact that is of course warmly welcomed by the aircraft designer.

However, liquid hydrogen also poses great challenges:

- For the same energy to be stored it needs four times as much volume than kerosene. It must be stored in well-insulated cylindrical or spherical tanks. As a consequence, the overall layout of the aircraft changes, which poses a wonderful challenge for our configurators, who are a bit tired of drawing aircraft that all look the same.
- There is a long list of subjects that still require R & D, e.g. fuel system layout, tank materials and insulation, system components, combustion chamber, heat exchanger, etc.

- A hydrogen-fuelled aircraft is very sensitive to the amount of fuel used. On a kerosene aircraft, we usually get the fuel volume 'for free', i.e. the wing box. In a hydrogen aircraft, an increase in fuel consumption causes an increase in wetted area and in weight, i.e. a strong snowball effect. Therefore, improvement of fuel efficiency by 'conventional' technical progress, e.g. by more efficient structures, is of even greater importance for hydrogen-fuelled aircraft than for kerosene-fuelled aircraft.

There is no doubt, however, that the technology is feasible and it is safe. The really big advantage of hydrogen is that it simply is the cleanest fuel you can imagine. It completely eliminates the emission of carbon dioxide, carbon monoxide, unburned hydrocarbons, soot and sulphuric acid. It allows for a significant reduction of the emission of nitrogen oxides in comparison to kerosene. So it will offer great benefits on a local/regional scale, even in case that it is produced from a fossil source like natural gas – it is easier to control emissions in a stationary chemical plant than in an aircraft engine. However, the advantage on the global scale, i.e. avoiding the emission of the greenhouse gas carbon dioxide, can only be achieved if the fuel is produced on a renewable energy basis.[1]

Why do we not introduce hydrogen right now? Again we have drivers and obstacles.

The actual need to reduce emission of greenhouse gases and the general public perception of our industry will drive us towards transition. Cars probably will set an example, which we will be expected to follow. The political objective to ensure long-term security of energy supply to Europe suggests increasing the use of renewable energies.

On the other hand, there is the general human trait of resisting innovations. ICAO is not really a fast-working organisation. There is the chicken-and-egg problem: aircraft and infrastructure. There is the need for operational proving of the new technology. 15 years from today to the entry-into-service of the first hydrogen-fuelled series aircraft are considered by some to be too short.

But the biggest problem is the economic side. Liquid hydrogen is an expensive fuel. It has no chance to compete with kerosene if the price of kerosene does not reflect its true cost to society.

Can we assume that politicians will do what is necessary to protect the environment effectively and establish a 'level playing field' for clean fuels vs. those spoiling the atmosphere? This may only happen under the pressure of climate changes becoming more and more obvious. It may happen late, but we had better prepare ourselves.

---

[1] The increased emission of gaseous water is of no importance if we stay at today's typical subsonic flight levels. According to computer simulations, the more frequent formation of contrails will be balanced by their lower optical thickness due to the lack of condensation nuclei in the exhaust jet. Contrails are a local and temporal phenomenon anyway, which can be largely avoided by proper selection of flight path/flight level.

We all carry responsibility for the well-being of our own industry, but equally to that of society. Hydrogen promises to be a way – perhaps *the* way – to approach our vision, i.e.:

> *To achieve long-term continuing growth of civil aviation until every man and woman on earth can fly as often and as far as they want and, when doing so, do not harm other human beings or the environment.*

As long as we do not know of anything better, we have no right to ignore the possibility of flying using hydrogen. In view of its great promise, we the engineers should rather accept with pleasure the challenge to develop the technology.

Response 4

Dilemmas and how to make a difference

C.A.M. (Kees) de Koning

Fokker Aerostructures B.V.

In only 100 years the aircraft industry has evolved from a pioneering start, with only very few people involved, to a mature industry with a yearly turnover of hundreds of billions of Euros and with an impact on almost everybody in today's 'New Atlantis' and beyond. A big impact when all goes well, but we notice this impact more when accidents happen or even worse when aircraft are being used as weapons or bombs. The tragic and unbelievable crashes in America on September 11, 2001 have rudely awakened us.

Flying is nowadays a very complex system of systems, far more complex than Francis Bacon could imagine with his extrapolation of 17th-century science and society. More complex and at the same time less perfect. Indeed, he would be disappointed. Here is a first lesson for engineers: 'Keep things as simple as possible!'

In the early decades of flying we saw pioneers, i.e. real entrepreneurs, inventors, scientists and businessmen driven by a fascination of flying or the quest for wealth and success. They invested their time and money to push flying forward, and soon after the first successful demonstrations the demand for flying quite literally gave them wings. Anthony Fokker was one of them and founded the company I represent today, see Figure 1. However, the pioneers have long gone and nowadays aircraft manufacturing is a very big global industry, employing hundreds of thousands of people and requiring vast capital investments.

An aircraft is a flying compromise, i.e. every aeronautical engineer learns the trade-offs between aerodynamics, structure, systems, manufacturing and maintenance requirements. On another level, an aircraft is always a compromise between economics, passenger comfort and safety, or in terms of performance the trade-off between speed, range, payload and economics. Above approximately 50 seats, in my view, the largely autonomous development resulted in the generation of aircraft, airworthiness regulations and infrastructure of today. This generation of aircraft is known as the Boeing 707 generation. The Boeing 707 flew first in 1954 and many are still commercially operational. Aircraft developed after the Boeing 707 are based on the same concept and fly at roughly the same speed. Of course, within the concept numerous refinements and further development took place, such as the use of composite materials, the introduction of fly-by-wire, the digital cockpit and, probably most importantly, the introduction of high by-pass engines with significantly better fuel efficiency. The Airbus A380 currently under development is the newest of this generation of aircraft. Here we face a first dilemma: 'Does the aircraft industry have to continue along this line and try to improve the current concept, or does the currently available technology enable us to start developing a family based on a new concept?' Boeing seems to think the latter at the moment considering its 'Sonic Cruiser', see Figure 2.

The Sonic Cruiser immediately leads to the topic of environmental impact, since as with almost all human activity flying has an impact on the environment. Although fuel consumption per passenger-kilometre is similar to cars and aircraft are among the capital goods with the longest operational life, emissions and noise are less and less accepted by the general public. In this, society faces a dilemma, i.e. the majority of the Western world population happily embraces the advantages of travelling by aircraft for both business and holiday purposes. However, at which price for the environment? Professor Lintsen gives the example of Concorde, where environmental concern stopped wider use of the aircraft. However, if we look at air travel on shorter regional routes, society accepts regional jets without any complaint instead of the much more efficient and quieter turboprops. Marginal gains in travel time, perceived safety problems of turboprops after some accidents, but in my view mainly the sexier image of jets win over the environment. The Sonic Cruiser undoubtedly trades environmental impact for higher speed and passenger appeal. Will (future) regulations allow such an aircraft to fly anywhere it could go?

Figure 1: Anthony Fokker and his first aircraft, the 'Spin'.

Figure 2: Artist's impression of Boeing's design for the 'Sonic Cruiser'.

This brings me to politics. As the military use of aircraft has become very important, the major powers in the world will most likely remain great supporters of aerospace and technology development. Not at any cost, but they will remain a major driver of technology development with its spin-off to commercial aircraft. Together with the enormous economical importance of flying and the high-tech image, governments will remain tempted to support their own industry. Not only do we see direct or indirect financial support, but biased regulations are also commonplace. Without this support the aerospace industry would look very different, and at the same time this industry has become dependent on government policies to a great extent. As government policies shift with times, this dependency is not at all comfortable, which poses another dilemma: 'Should the aerospace industry accept government dependency as a fact, or should they steer their own course?' A consolidation process took place in both Europe and the USA driven by demands for greater (cost-)effectiveness by governments.

'Big is beautiful!', seems to be today's aerospace industry motto. Five hundred years ago Leonardo da Vinci wrote down his ideas of a flying machine. One brilliant man was able to understand almost all the knowledge of his time and to become an expert in many fields, i.e. technology, arts, philosophy and more. At the time of the flying machine's pioneers, one man was able to oversee almost all relevant knowledge for building better aircraft. In the late twenties a small team of very capable but general engineers could build an aeroplane, as demonstrated by a Stork company who designed the *Werkspoor Jumbo* at the request of KLM, see Figure 3. Only one was built, which successfully transported gold between European capitals for over 10 years until the Germans destroyed it in 1940.

Figure 3: Artist's impression of the sole Werkspoor Jumbo ever to take to the air.

Now the design and development of a new aeroplane requires thousands of engineers, each of them a specialist in his own field and working together in very complex organisational structures. Only very large companies can maintain and develop the required manpower, expertise and infrastructure. The development of a new aeroplane has become very costly, e.g. in case of the Airbus A380 more than 10 billion Euro. With this kind of investment 'try-and-see' can become a very costly affair. From a financial point of view the margins between failure and success are small.

Consolidation of the aerospace industry has resulted in only a few, very large companies. For instance, in the important large passenger aircraft segment only two integrators remain, i.e. Airbus and Boeing. A new competitor in this segment is virtually impossible in view of the enormous and long-term investments required. In my opinion, sooner rather than later only two integrators will remain in other aerospace segments such as business jets, regional jets or aero-engines. In order to survive in the long run, decisions on new aircraft programmes and the new technology to be incorporated will have to be taken with great caution. It does not need much imagination to see the resulting dilemmas put before the decision makers in these companies.

Due to the required investment capital for new programmes and because the large integrators simply can not specialise in all required areas, the few remaining aircraft manufacturers will increasingly concentrate on their role as 'integrator'. They assemble a team of 'first tier' partners who share the financial risk by investing in integrators' programmes and provide resources and expertise for specific parts of the aircraft. Here I see another dilemma, i.e. integrators, or rather their people, still only feel comfortable if they themselves control the required expertise and yet they must team up and often leave others in control. Companies and people will have to learn the do's and don'ts to make those very long-term partnerships work.

In just such a role, Fokker teamed up with Airbus for the design and manufacturing of Glare fuselage panels for the A380, see Figure 4. The blue areas in Figure 5 are planned to be Glare panels, some 400 m^2 in total. 20 years ago, when I was studying at the Delft University of Technology, Professor Boud Vogelesang asked me to go to Fokker to investigate the fabrication of Arall stringers. Now, 20 years later, the combined efforts and expertise of the Delft University of Technology, the National Aerospace Laboratory NLR, Akzo, Alcoa, FMLC, Airbus and Fokker have resulted in the application of Glare technology in the Airbus A380. A team of so many parties and a 20-year lead-time are typical for a new technology application in aircraft. The Netherlands, with its excellent technological R & D infrastructure, can team up with aircraft integrators in its various fields of competence and assist in building better flying machines. Glare has been a very successful development. However, the dilemmas remain. Many technology avenues lay ahead, and we must investigate which of these will lead to success.

Figure 4: Concept Glare fuselage panel by Fokker Aerostructures.

I would like to add one more conclusion to the six defined by Professor Lintsen, i.e.: 'Individuals can make a difference!' Of course, we saw that Boeing and Douglas should be included with the likes of Leonardo da Vinci and pioneers like Anthony Fokker. However, even today, with the vast extent and complexity of aerospace technology, one man can still make a difference. e.g.:

- The quality and success of the Boeing 777 is due to a great extent to how Phil Condit was leading his 777 team in an entirely new organisation.
- Without the strong belief and tireless drive of Jürgen Thomas, Airbus may not have launched the A380.
- Closer to home, without Boud Vogelesang we would not be here today (ed.: September 24, 2001 – day 1 of the *Glare - The New Material for Aircraft* Conference – in the Aula Conference Centre of the Delft University of Technology in Delft, the Netherlands) and there would be no large-scale Glare application in the Airbus A380.

I would like to thank Boud Vogelesang now for his enthusiasm, inspiration and never-fading belief in this kind of technology development. I strongly believe that such leaders will inspire and guide us in making the best possible use of new technology, and that will make a difference!

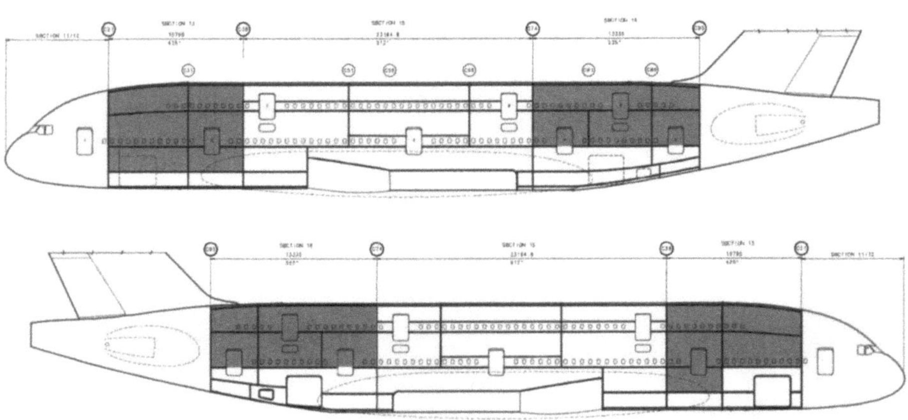

Figure 5: Glare application area on the Airbus A380 per September15, 2000.

Response 5

Daan Krook

Independent aviation consultant

The paper read by professor Lintsen has undoubtedly been met with great interest by all of you. I should like to summarise his six conclusions as follows:

1. The 20th century was the century of technology.
2. Development of technology can be viewed as autonomous, as a matter of choices or as unpredictable.
3. Complexity must be accepted and consequently so should openness and flexibility.
4. Optimism should be guided by pessimism.
5. Realisation requires time.
6. Quality is more important than quantity.

In these conclusions I missed one factor, which I think very often influences the technological development and certainly has influenced and still does influence the development and the application of Glare. This factor is the economic-political influence from governments and their agencies and from large industries.

As an example I would like to take the case of the Concorde, as mentioned by professor Lintsen. Although undoubtedly genuine environmental factors played an important role, as Professor Lintsen indicated, the fact that the Concorde was European was, in my opinion, an important factor that limited its operational approvals and therefore its success in the market. Also, in the case of aircraft noise in general, there is reason to doubt whether the rules and figures were indeed based upon impartial requirements or in actual fact along the lines of the capabilities of existing aircraft types that had already been sold in large numbers. More recently, burn-through factors for fuselages and evacuation times for aircraft are judged in the light of what is acceptable for existing aircraft instead of what is technologically possible. Another case is that of the bombproof containers, which are possible but not used.

My conclusion is that technology, without undue speed and optimism, is often not given the opportunity to be applied in the interests of the public. The reason I voice this opinion is because the development and application of Glare were also influenced by such economic-political factors. The material showed right from the beginning promise in many of the fields that have been problem areas for the aircraft industry, e.g. weight saving, fatigue, crack propagation, fire hazards, isolation, excessive wear and consequent maintenance, to name just a few. However, neither the government agencies, nor the metal industry, and for that matter nor most of the aircraft producers and airlines, showed active interest, whereas in my opinion, some famous accidents or losses of lives could have been prevented if the fuselages or cargo holds had been built out of Glare. By many, the product was shelved for future reference, if not secretly or openly opposed. Fortunately, however, this was with one exception, i.e. Airbus Industrie, whose interest turned out to be of major importance. Admittedly there was more support, but that came mainly from individuals in government, universities and some industries and not as a matter of policy.

The obvious question for all of us is of course what the future will look like. There is no doubt that the aviation industry will continue to grow and therewith the aircraft industry. Only a few percent of the world population has ever flown and the remainder will want to fly as soon as they are given the opportunity. Out of necessity, they will have to be transported by aircraft that are more environment-friendly and safer. These can only be achieved by making use of the most modern materials.

The past decades have shown that progress in development is helped tremendously by competition, which was virtually non-existent, at least in the larger segment, until the arrival of Airbus Industrie. This competition will also be found in the regulating agencies to a greater extent than was the case until the 1970s and they in turn will be forced by other government circles to make certain that the socially required air transport will be environmentally acceptable. There is no doubt that this will lead to the development of new aircraft types, even though the existing types still have a lot of life in them, technically speaking. This will also press for the use of new materials. Finally, special military requirements will also point in the same direction, i.e. better protection and performance by better design and materials.

Coming back to the conclusions of professor Lintsen, I should like to state that in the twentieth century, the age of technology, the development of Glare was a matter of choice, but maybe even more of belief, and that for many it was predictable but for some apparently a threat. It was complex for its proponents, but they remained open, flexible and above all optimistic. They had to take more time than they liked, but thereby provided quality and hopefully quantity as it takes its place in the future.

Day 2: DEVELOPMENT OF MATERIALS FOR AIRCRAFT DESIGN

Sponsored by: DELFT AEROSPACE
FACULTY OF AEROSPACE ENGINEERING

Keynote lecture

Materials and the development of aircraft: Wood - aluminium - composites

Eric M. Schatzberg

Department of the History of Science
University of Wisconsin-Madison

The utility of history

According to Hegel, the owl of Minerva only flies at dusk. By this Hegel meant that the working of reason in human affairs only becomes apparent at the end of history. Although Hegel formulated this aphorism in opposition to normative theories of the state, it also sheds light on technological change. Minerva, the Roman version of Athena, was not only the goddess of wisdom and war, but also of many practical arts. The Greeks gave her credit for numerous inventions, including the flute, clay pot, plough, and ship.[1] If we apply Hegel's aphorism to technology, it suggests that the cunning of reason only becomes clear at the end of the innovation process, not during the messy complexity of technological change.

This understanding of technology has two implications. The first applies to what Bruno Latour [3] calls 'technology in the making', that is, the innovation process.[2] Even people who create scientifically sophisticated technologies do not experience the process as entirely rational. As most design engineers readily acknowledge, design and innovation can not be reduced to a science. In the creative struggle to shape a new technology, reason's light does not shine brightly enough to penetrate the fog of uncertainty obscuring the best path to success.

[1] What Hegel actually wrote, was: '*Wenn die Philosophie ihr Grau in Grau malt, dann ist eine Gestalt des Lebens alt geworden, und mit Grau in Grau läßt sie sich nicht verjüngen, sondern nur erkennen; die Eule der Minerva beginnt erst mit der einbrechenden Dämmerung ihren Flug*', [1]. On Minerva/Athena, see Graves [2].

[2] Latour actually refers to 'science in the making', see Latour [3] (p.4), but he sees no fundamental difference between science and technology, the making of facts and artefacts.

The second implication concerns the utility of philosophical and historical understanding for people engaged in technological innovation. Hegel intended his statement as a warning against philosophers who sought to construct prescriptive theories of the state, that is, to tell the state what it ought to be [4]. With regard to politics, this viewpoint is profoundly conservative. With regard to the history of technology, though, this caveat is worth heeding. Neither historical nor philosophical reflection can tell innovators what to do in specific circumstances.

These limitations do not mean that history is irrelevant to engineers and scientists involved in innovation. Successful innovation is the product of practical wisdom as much as scientific theory. The essence of practical wisdom is the art of judgement, knowledge of how to make correct choices in specific contexts. Aristotle insisted that practical wisdom could not be taught directly, but could only be learned through a proper upbringing, that is, through socialisation within a culture. I do not entirely agree with Aristotle. While historical reflection can not replace practical wisdom gained through socialisation, it can supplement the socialisation process.[3] Viewing past innovations from the perspective of Minerva's owl can prepare technologists to deal with the practical issues that arise in creating a new technology.

Keeping these limitations in mind, I believe that history suggests three general requirements for a successful technological innovation. First, the technology must succeed as a material object or process, or more precisely, as a system of related artefacts and material processes. From this perspective success is defined as the reliable manipulation and transformation of energy or force to achieve desired goals [6]. Second, the technology must succeed in the realm of human practices, that is, routinised patterns of human interaction with the material world. Technological practices include all the tacit knowledge and standard procedures involved in the design, production and use of a technology. New technologies invariably require changes in practices, particularly among users, because no new technology is completely ‘plug-compatible’. New materials in particular require extensive changes not only in the practices of design and production, but also in maintenance and repair procedures.[4]

There is a third requirement, however, that does not get as much attention as the first two, especially by technical people. A new technology must also succeed in the realm of symbolic culture, as a thing endowed with human meaning. In other words, a successful innovation must make sense to the social groups with power to influence its creation and adoption.[5] A successful innovation requires more than just

---

[3] On the connection between Aristotle's concept of phronesis and the exercise of judgement in relation to science, see Bernstein [5].

[4] On scientific practice, see Polanyi [7]. Practice has been a topic of considerable recent interest in science studies, but technological practice has received little separate attention. For a recent discussion, see [8].

[5] John Staudenmaier has termed this approach the ‘cultural construction of technology’ in his review essay ‘Recent Trends in the History of Technology’ [9]. The concept was implicit in early work in the social construction of technology, particularly in Trevor J. Pinch and Wiebe E. Bijker's seminal

convincing customers that the innovation will meet practical needs. New technologies are invariably like newborn babes, of practical value only after a long period of nurturing. Support for technology in the making can not, therefore, be based on its existing practical benefits, because if such benefits were already present there would be no need to fund R & D. Instead, the proponents of a new technology must create an imagined future in which the technology plays a key role, and they must sell this imagined future to people whose support is required for the innovation to achieve maturity.

This paper focuses on the third requirement, i.e. success in the realm of symbolic culture. Recent work in the economics of technological change highlights the centrality of expectations in the choice between competing technologies. Expectations are shaped, I argue, as much by the symbolic meanings of materials as by their technical promise. The struggle between metal and wood during the 1920s provides a poignant example of the symbolic shaping of technological choice. Symbolic meanings have also shaped the choice between metals and composites since World War II, although in a more subtle manner. The developers of new materials like Glare can benefit, I believe, by paying attention to the symbolic significance of their product as well as its physical advantages.

Path dependence theory and the symbolic shaping of technology

Symbolic culture receives little attention in most discussions of innovation. Yet recent work in the economics of technological change provides strong (though implicit) support for taking seriously the role of symbolic culture in technological change. I am referring to path dependence theory, which is most strongly associated with the economist W. Brian Arthur.

At the heart of Arthur's theory are the non-linear relationships that exist between inputs and outputs in what he terms knowledge-intensive industries. In traditional resource-intensive industries, unit costs tend to increase with scale due to resource constraints. In knowledge-based industries, however, unit costs tend to decline with increasing size, scope, and production experience. A key element in this non-linear relationship is the learning curve, which was first quantified in the production of aircraft during the 1930s [11, 12, 13, 14].[6]

In these markets characterised by increasing returns to scale, the standard assumptions of neo-classical economics break down. Such markets lack the invisible hand that automatically steers firms toward the technology that will maximise profits

paper 'The Social Construction of Facts and Artifacts: Or How the Sociology of Science and the Sociology of Technology Might Benefit Each Other' [10].

[6] It was the prominent aeronautical engineer T.P. Wright who published the first empirical study of what was later called the learning curve, although Wright insisted that the phenomenon was 'well-known' among experts in manufacturing efficiency [15].

relative to the prices of labour, capital and materials. Instead, technologies compete in a dynamic process that tends to 'lock in' one technology at the expense of its competitors. Market mechanisms do not insure the victory of the technology with the best long-run potential, as demonstrated by the ubiquity of the QWERTY keyboard in English-speaking countries.[7] Furthermore, the temporal sequence of change plays a key role in determining which technology succeeds. Small early events can have major long-term consequences; being first to market can sometimes be more important than having the best technology. In other words, the outcome depends on the path taken to get there. Finally, models of these markets suggest that expectations of success tend to be self-fulfilling. Because the technology that eventually achieves lock-in will have the lowest actual if not potential cost, rational actors will tend to choose the technology that they believe most people will prefer, even if they believe this technology to be sub-optimal.[8]

Although path dependence theory has been subject to substantial criticism by neo-classical economists, it makes considerable sense to historians of technology and, I believe, also to technological innovators. All new technologies are knowledge-intensive in Arthur's sense, because the cost of the first item produced is dominated by the fixed cost of R & D. All innovators need support to move their innovation far enough down the learning curve to enable it to compete against established technologies. Successful inventors understand this need, and they tend to be good promoters as well as skilled technicians. Thomas Edison was especially adept at managing expectations surrounding his work. For example, Edison announced to the press that he had solved the problem of the incandescent light in September 1878, when in fact he had only come up with the germ of an idea, one that later proved unworkable. Nevertheless, this claim created an expectation that Edison would be the first to market with a workable system, which helped Edison secure the financial backing that was necessary to his success [19]. Likewise, a failure to create expectations of success can doom a technically promising innovation. Most engineers know of excellent products that failed due to poor marketing rather than technical flaws.

Path dependence theory has helped bring expectations to the foreground of technological change, but it does not explain where expectations come from. In part, expectations are driven by scientific understanding of the inherent potential of competing technologies; such understanding stimulated interest in fibre-reinforced composites, for example. Scientific knowledge can not, however, completely remove

---

[7] On QWERTY as an example of path dependence, see David [16]. The example of the sub-optimality of the QWERTY keyboard has been attacked by S.J. Liebowitz and Stephen E. Margolis [17]. Their argument fails to establish the optimality of the QWERTY keyboard, but suggests that the inferiority of the QWERTY keyboard has been exaggerated. Nevertheless, even if the QWERTY keyboard only produces a small loss in efficiency compared to alternative keyboard layouts, the economic costs are still staggering. Perhaps a stronger case is the dominance of DOS, which Arthur cites as a clear example of an inferior technology achieving lock-in. There were a number of better-developed microcomputer operating systems at the time, but IBM chose Microsoft with little consideration of alternatives, such as porting Unix to the PC [18].

[8] David [16] is particularly clear on these points.

the uncertainty of technological choices. Science too must heed Minerva's owl; science can not tell us if a particular theoretical possibility can ever be exploited practically. Controlled nuclear fission was achieved only four years after its first detection on a laboratory scale; controlled thermonuclear fusion still remains a distant goal despite half a century of large-scale research and development.[9] Investment in a new technology always requires a leap of faith.

And where does this faith come from? Ultimately it comes from the significance of a technology within the complex webs of symbolic meanings that constitute the cognitive part of a culture. Proponents of a particular technology draw on specific associations that connect their technology to powerful cultural symbols, most importantly the symbolism of technological progress. Culturally speaking, the most powerful symbolic association for a new technology is its metaphoric designation as the 'wave of the future'. This symbolic link between a technology and modernity can serve as a powerful material force, convincing both investors and users to nurture the new technology through its problematic childhood and troubled adolescence into a mature innovation. In this way, symbolically shaped expectations of success tend to become self-fulfilling.[10]

Expectations alone, however, can not insure the success of an innovation. Symbolic meanings constitute only one of the three realms in which a new technology must succeed, and can not substitute for success in the realms of material artefacts and human practices. Cost and performance remain critically important in the choice between competing technologies. Yet, because costs and performance can not be accurately predicted before investing in the development of a new technology, symbolically shaped expectations can themselves influence which technologies are chosen for development.

Choice of materials in aeroplane design: technical indeterminacy

The shift from wood to metal aeroplanes nicely illustrates how symbolic meanings shape expectations, and how expectations shape the innovation process, most importantly by influencing the allocation of resources for research and development. One can view the shift from wood to metal as an example of Hegel's cunning of reason. The end-point of the process was the fully streamlined, stressed-skin aluminium alloy airframe of the mid-1930s, exemplified by the Douglas DC-1 and 2, see Figure 1. This type of structure was a technical triumph, providing excellent weight efficiency, aerodynamics, and durability. Most non-technical people today assume that the choice of metal for these structures was dictated by engineering criteria. Yet I argue that the shift to metal was driven as much by symbolically

[9] Fission was discovered by Hahn, Strassman and Meitner in December 1938; Enrico Fermi's first full-scale reactor went critical on December 2, 1942; see Rhodes [20]. For a survey of American efforts on controlled fusion, see Rowberg [21].

[10] On self-fulfilling expectations in technology, see MacKenzie [22].

shaped expectations as by metal's rigorously demonstrable advantages. The all-metal aeroplane was embraced by the aeronautical community in Europe and the United States well before the success of the DC-2 and its kin. And it was in part because of this embrace, this expectation that metal would inevitably supplant wood, that the metal aeroplane achieved the success that it did.

I am not arguing that wood was superior to metal for aeroplane structures, or even that it would have been superior if it had received comparable R & D support. Probably more research could have helped wood retain a place in small aeroplanes, but I for one am very happy to fly in today's all-metal airliners. Nor am I arguing that supporters of metal were acting irrationally, or at least any less rationally than supporters of wood. From the perspective of Minerva's owl, at least, the choice of metal seems to have been quite rational. But from the perspective of the interwar aeronautical community, the choice of metal involved a leap of faith that was not and could not have been fully justified by technical criteria. Such leaps of faith almost always accompany decisions to embark on the development of a new technology. As a consequence of this leap of faith, humankind obtained a technology that helped make the dream of flight a reality for millions of people. Yet even with the undisputed success of the streamlined all-metal stressed-skin airframe, we can not be certain that the optimal path was the metallic one. Much was gained by focusing effort on metal structures, but something was lost as well – the potential contribution from the non-metallic path, whose promise is only now being fully explored through the development of fibre-reinforced plastics.

Figure 1: Douglas DC-1 prototype, March 6, 1934 [23].

Every aeronautical engineer knows well the key technical problem posed by the choice of materials in aircraft design. It can be summed up in one word – weight. Weight engineering plays a more central role in aerospace structures than in any other branch of technology. The American aeronautical engineer William F. Durand expressed this problem quite clearly in the Sixth Wilbur Wright Memorial Lecture before the Royal Aeronautical Society in 1918: 'Broadly speaking, the fundamental problem in all airplane construction is adequate strength or function on minimum weight', see Durand [24] (33). Eighty-three subsequent years of aeroplane design have not diminished the importance of weight in the choice of aeroplane materials.

Aeronautical engineers also know that the relationship between the properties of materials and the weight of a complete structure is not simple. This relationship depends on the geometry of the structure and the forces it must bear. In the design of some parts weight is inversely proportional to density, in others to density squared or cubed. Some designs are governed by ultimate strength, some by yield strength, some by fatigue strength and some by Young's modulus.[11] Most non-metallic materials are highly anisotropic, further multiplying the variables to consider. Weight efficiency can not be assessed by substituting materials in existing structures; each material demands its own structural design in order to take advantage of its specific properties.[12]

But even if weight is the primary criterion of aeroplane design, it is not the only one. All engineering is about compromise, and weight must always give some ground to other criteria, such as durability or ease of manufacturing. Most fundamentally, these compromises embody a trade-off between cost and performance.

Yet the connections between cost, performance and the choice of materials are extremely complex. I use the term 'technical indeterminacy' to describe this uncertain relationship between technical criteria and the choice of materials. All new technologies face similar problems of technical indeterminacy. This indeterminacy arises because the criteria of design inevitably conflict, requiring compromises between competing goals. Every designer must make choices, whether between first cost and durability or between power and efficiency. As Curtiss-Wright chief engineer T.P. Wright noted in 1929: 'It sometimes seems that there exists no element of design which does not conflict directly with every other element', [29].[13] Furthermore, there is no rational calculus for balancing these competing criteria, just as there can be no ultimate rules for applying rules. In practice, technical choice always involves reasoned judgement as well as rational calculation.[14]

---

[11] I find J.E. Gordon especially insightful in this regard, see Gordon [25, 26, 27].

[12] Early aeronautical engineers were quite aware of the difficulty of comparing materials apart from the structures designed to take full advantage of their properties, see Durand,[24] (34) and Warner [28].

[13] This view of aeroplane design as a compromise between conflicting goals was widely shared among aeronautical engineers in this period in Europe and America [30, 31, 32, 33].

[14] On this point see esp. Pye [6] (70). There are actually several independent ways to argue for the inevitability of technical indeterminacy, see Schatzberg [34] (17-18).

Technical indeterminacy is not an absolute condition but rather a matter of degree, experienced more by new technologies in flux than by stable products. Design choices at any given time always face a certain degree of technical indeterminacy, but in practice engineering judgement can usually resolve such indeterminacy with little controversy, for example in a choice between threaded or riveted fasteners.

More significant is the indeterminacy involved in choosing a path leading to a major new technology, that is, choices about R & D. Such choices involve deep uncertainties about the future characteristics of the technology under development and the future human practices necessary for the technology to succeed. This uncertainty is particularly acute with regard to costs of the finished product; engineers have much more confidence in achieving their performance goals than their cost estimates. Uncertainty with regard to costs demonstrates that economic criteria provide no more than a rough guide to technological change at the level of the firm or research institute, just as path dependence theory suggests that markets do not choose between competing technologies in a deterministic way. Early in the innovation process, beliefs about future costs are more significant than actual costs in shaping investments in R & D.

Research and development serves to reduce indeterminacy by creating knowledge and practices that shift the choice decisively in one direction or another. In the case of aeroplane materials, this shift occurred by about the mid-1930s, when metal's superiority became moderately well established, at least for high-performance aeroplanes. Yet that choice was certainly not clear in the early 1920s, when the aeronautical community decisively embraced the development of metal aeroplanes. The rationality of a particular R & D path can only be determined after embarking down that path. The owl of Minerva only flies at dusk.

Wood versus metal during the 1920s

Aeronautical engineers at the end of World War I faced just such a period of technical indeterminacy concerning the future of aeroplane materials. During the war, aeroplane design and construction had emerged from the world of self-taught designers building a few aeroplanes in small workshops. By the end of the war, the major powers had produced approximately 170,000 aeroplanes [35], but more importantly the combatants had established major technical centres to undertake what we would now term research and development on all facets of flight technology. Aeronautical engineering had become a recognised speciality involving sub-fields in aerodynamics, structures, materials and engines.[15] With the end of the

[15] Although Mark Dierikx (in this volume) is certainly correct that the influx of capital transformed the American aeroplane industry in the late 1920s, this transformation did not involve a fundamental shift in aeroplane design from an 'empirical' to a 'scientific' approach, as Dierikx suggests. Rather, this shift had already occurred during and soon after World War I. By 1920, every major power had established research facilities employing thousands of scientists, engineers and technicians. In the United States, for example, this R & D infrastructure included the National Advisory Committee for

war, aeroplane production collapsed, but technical change continued at a rapid pace. Aeronautical engineers had to choose among a variety of promising technical paths for aeroplane design. Nowhere was this choice more stark than in airframe materials.

The vast majority of aeroplanes built during the war had fabric-covered wood structures. A small but significant number used the welded steel tube fuselage pioneered by the Dutch aeroplane manufacturer Anthony Fokker. But another development was more potent symbolically for post-war debates; namely the design of all-metal aeroplane structures made predominantly from duralumin, the first precipitation-hardened aluminium alloy. By the end of the war the Germans had produced a few hundred serviceable aeroplanes with these all-metal structures, most significantly the Junkers J4 armoured ground attack biplane, see Schatzberg [34] (chap. 2).[16]

Simply by their existence, these all-metal designs raised the question of which path to choose for aeroplane materials. All-metal construction had its impassioned supporters who insisted that metal in general and aluminium alloys in particular offered the most promise of any aeroplane material. Such advocacy is the norm for new technologies, and supporters of metal worked hard to build an imagined future in which the triumph of metal would appear inevitable. Advocates of metal had tremendous success in gaining support for this imagined future, but this support was not obtained solely on the basis of technical arguments.

My argument for the indeterminacy of the choice between wood and metal is counterintuitive even for aeronautical engineers, given the tremendous success of

---

Aeronautics and its Langley laboratory, the Army Air Service engineering centre at McCook Field (predecessor to Wright-Patterson), the Navy Bureau of Aeronautics in Washington and the Naval Aircraft Factory in Philadelphia, with significant additional research performed by the National Bureau of Standards and the Forest Products Laboratory. Furthermore, the new Aeronautical Engineering programme at the Massachusetts Institute of Technology provided graduate training to dozens of engineers in the early 1920s. Even when aircraft companies continued to be owned by self-taught entrepreneurs, these firms hired highly-trained engineers who took full advantage of the extensive research reports published by the NACA and other agencies. Furthermore, the armed services were quite willing to spend huge sums on engineering new designs; William Stout, for example, received roughly $200,000 from the Navy in the early 1920s to develop his unsuccessful ST-1 all-metal torpedo bomber; see Schatzberg [34] (64-66, 86, 90). On aeronautical education in the United States, see Schatzberg [36].

[16] Contrary to Dierikx' portrayal of Hugo Junkers in this volume, Junkers' wartime metal aeroplane work has all the marks of technological enthusiasm, as does his earlier collaboration with Hans Reissner on the '*Ente*'. Certainly no present-day structural engineer would take seriously the idea of an all-metal wing for a small aeroplane like the '*Ente*', especially using sheet iron or pre-duralumin wrought aluminium alloys. The same technical judgement applies to Junkers' J1 of 1915, which also used sheet iron as a wing covering. This aeroplane had a wing loading of about 3.4 lb/ft^2 (37 kg/m^2), far too low to utilise the maximum strength of the material, especially given practical minimum sheet gauges. Furthermore, the J1 did not come close to meeting the minimum performance requirements for a combat aeroplane, as Junkers himself admitted later. The 'scientific' Junkers devoted tremendous resources to developing all-metal aeroplanes for the German military during the war, yet produced only one modestly successful aeroplane, in sharp contrast to Fokker, whose 'empirical' approach yielded over a thousand fighters that competed on equal terms with the best British and French aeroplanes, see Schatzberg [34] (24-26).

stressed-skin aluminium alloy structures in airframe design. I therefore want to review briefly the technical situation with regard to aeroplane materials in the early 1920s.[17]

The most important issue faced by aeroplane designers was weight efficiency. As I noted above, comparing the weight efficiency of materials is difficult. Ultimate strength was the main concern of aeronautical engineers in the 1920s. In terms of ultimate tensile strength, the best woods, typically spruce, had superior weight efficiency to all but the strongest speciality steels. Tests of materials in compression gave a clear though not huge advantage to duralumin and high-strength steel. Yet when it came to building actual metal wings, most designers found it very difficult to achieve weights comparable to those of the best wood structures.

The reason for this difficulty was quite simple – compressive instability, typically through local buckling in the thin flanges of metal beams, see Figure 2. Wood was also subject to buckling failures, but since the buckling strength of a flat plate varies inversely with the cube of density, wood held a considerable advantage, with spruce plywood weighing about 60 percent less than aluminium for equal buckling strength.[18] Surprisingly, buckling was almost never raised as a criterion for comparing wood and metal in the 1920s. Yet at the wing loadings common in the 1920s, there was no simple way for designers to concentrate enough material to avoid the danger of buckling without a significant weight penalty. Early American metal designs proved disastrous in this regard; the Army's first metal bomber, delivered in 1921, was too overweight to carry any bombs. Even supporters of metal construction admitted that metal wings were in general heavier than wood wings, although this disadvantage decreased as designers gained more experience with metal structures.

But when engineers did produce metal wings comparable to wood in weight, the complex systems of reinforcement required to avoid buckling invariably increased manufacturing costs, especially after the adoption of stressed-skin designs in the early 1930s. Metal aeroplane prototypes could cost an order of magnitude more than comparable wooden prototypes. In the late 1930s, after most American aeroplane manufacturers had a full decade of experience with metal aeroplane production, metal aeroplanes still cost twice as much per airframe pound as their wood-and-fabric predecessors, see Bright [37] and Schatzberg [34] (52-54).[19]

---

[17] This discussion is based on Schatzberg [34] (44-56), unless otherwise noted.

[18] Using specific gravity of 0.5 for spruce plywood and Young's modulus of $1 \cdot 10^6$ psi (6.9 GPa).

[19] The idea that metal only became practical with quantity production of aeroplanes in the 1930s is one of the most persistent falsehoods in the technical history of aviation. Certainly larger production runs justified greater development expenditure, but this advantage applied both to metal and wood aeroplanes. Furthermore, the air transport market remained very small before World War II, especially for multi-engine passenger aeroplanes, of which the United States produced only 53 in 1938. The main aeroplanes to benefit from quantity production in the interwar period were small single-engine models; precisely the type that remained dominated by wood and fabric wing construction.

Figure 2: Wood and metal wing spars tested at Wright Field, 1927; note buckling failures in top flange of second and third spar from left [38].

In terms of weight and costs, all-metal construction provided no clear advantages over wood in the 1920s. However, practical experience did demonstrate that metal could achieve comparable weight efficiency, and proponents explained away high initial costs by promising substantial savings in quantity production. Since metal could not demonstrate superiority in terms of weight and initial cost, its supporters turned to two other key technical criteria, safety and durability.

Advocates of metal aeroplanes insisted that metal provided great protection against aircraft fires, which were all too common with the gasoline-fuelled piston engines of the 1920s. This claim might have had some validity if steel had triumphed instead of aluminium. But aluminium's low melting point made it almost worthless as a fire barrier. Although I know of no tests comparing the fire resistance of plywood and aluminium sheet in the 1920s, tests in the late 1930s demonstrated a tremendous advantage for Bakelite-glued plywood – by the way, Glare has a similar advantage over aluminium. Practical experience with aluminium alloy aeroplanes in the 1920s showed them to have no advantage over wood in fire safety; in fact the US Air Mail abandoned its early experiment with all-metal Junkers transports in 1921 after a series of fatal crashes linked to fuel fires, see Schatzberg [34] (45).[20]

Durability provided perhaps the strongest argument in favour of metal, one repeatedly invoked by its advocates. Hugo Junkers was particularly forceful on this subject, claiming that wood suffered from decay, dimensional instability due to changes in moisture, deterioration of glued joints, and even 'attack ... by insects'. Wood indeed had serious durability problems, but metals did as well. Duralumin-type alloys were especially susceptible to inter-crystalline corrosion, see Figure 3,

[20] On Glare, see Vlot, et al [39].

which proved particularly frightening because of the lack of surface indications. By the end of the 1920s, corrosion problems had cost metal considerable support among aeronautical engineers. As one manufacturer remarked: 'For durability and dependability I'll have my all-metal airplanes made of wood', [40]. Concerted efforts by government researchers and the aluminium industry produced a solution to this problem, most notably through the use of pure aluminium cladding. Comparable efforts were not undertaken to solve the durability problems of wood aeroplanes, even though synthetic resin adhesives promised substantial improvements in the durability of glued joints, see Schatzberg [34] (54-56, 92-95, 174-190).

In the long run, the hygroscopic nature of cellulose fibres would probably have led to the development of alternatives to wood as an aeroplane material. Nevertheless, in the 1920s there was no quantitative evidence that metal aircraft provided significant maintenance advantages over wood. Some evidence was forthcoming in the 1930s, but these results were the product of engineering development and learning curve effects, and therefore can not provide a causal explanation for earlier support for metal. In other words, even though the belief in the potential superior durability of metals was reasonable, it still required substantial technological development to make this potential a reality. Until this superiority could be manifest, the choice between wood and metal remained indeterminate.

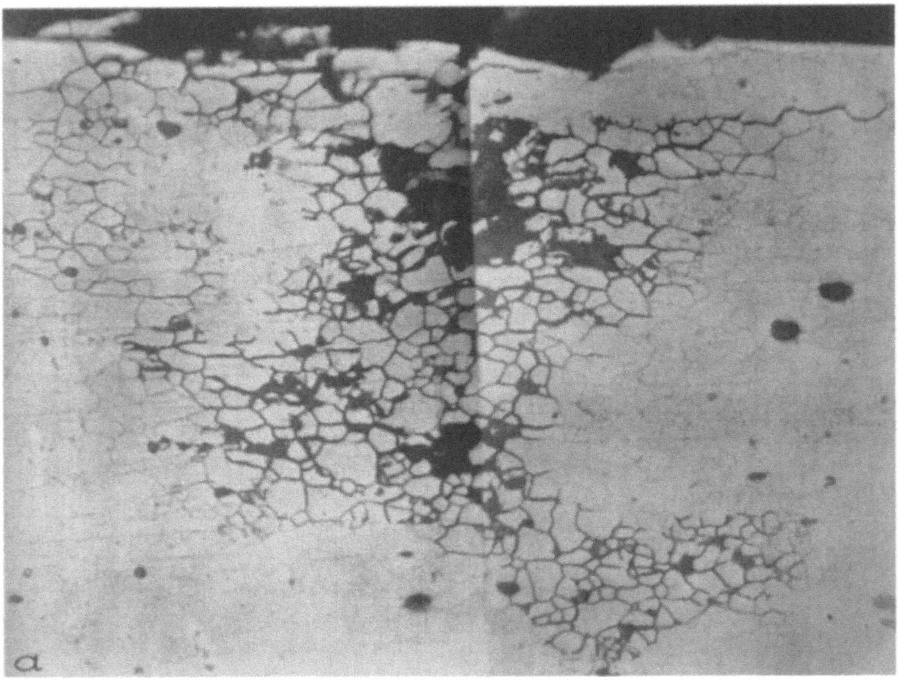

Figure 3: Inter-crystalline corrosion in duralumin-type alloy [41].

This portrait of technical indeterminacy in the choice between wood and metal during the 1920s is based on published sources and internal documents available to the organisations with power to shape the technical development of aviation, namely the aeroplane manufacturers and government agencies, especially the armed forces. These documents show clearly that neither wood nor metal could demonstrate any overall advantage as an aeroplane material in the 1920s, at least in the United States. In terms of weight, theoretical considerations gave no clear advantage to either material. In practice, metal wings on average weighed more than comparable wood wings; though by the end of the decade this difference had declined. The first cost of metal aeroplanes remained significantly greater, especially for all-metal stressed-skin designs. Metal aeroplanes failed to demonstrate any safety advantages, particularly with regard to fire. Information on maintenance costs remained anecdotal and indecisive, with no quantitative comparisons of maintenance costs of aeroplanes in comparable service conditions.

Explaining the choice: symbolic meanings

Given this uncertainty, why did the aeronautical community, both in the US and Europe, provide such strong support for the development of metal aeroplanes in the 1920s? Perhaps aeronautical engineers in the 1920s understood that metal structures would be preferable for the large, high-speed aircraft that emerged in World War II and beyond. It is rather difficult to imagine a Boeing 747 with a wood structure, or even a Boeing B-17.[21] The trend to larger and faster aeroplanes required thicker wing and fuselage skins, which reduced metal's disadvantage in buckling strength. Also, as wing loads increased, stresses in structural members eventually had to increase as well, because low-density, lightly-stressed materials like wood could not support the required loads within the confined spaces of wing structures. De Havilland engineers faced this problem in 1944 when they considered strengthening the wooden wings of the Mosquito to reduce structural failures in high-speed manoeuvres; unfortunately there was not enough room to add material to the wing spars.[22]

This kind of retrospective explanation of support for metal might appear rational to the owl of Minerva. There is no evidence, however, that such reasoning existed in the heads of the historical actors. Although advocates of metal like Hugo Junkers claimed that metal was necessary for large aeroplanes, they did not support this argument technically. Nor have I found any evidence that aeronautical engineers made a connection between wing loads and the choice of materials during the entire interwar period. In any case, few aeronautical engineers in the 1920s anticipated the

[21] Actually, this difficulty reflects on the poverty of our historical imagination; a number of engineers did propose large aeroplanes built from resin-impregnated plywood, most importantly Howard Hughes. Hughes failed with his 400,000 lb gross weight HK-1 (the 'Spruce Goose'), but a resin/plywood aeroplane comparable to a B-17G at 55,000 lb (25,000 kg) normal GW could very well have been more successful; see Schatzberg [34] (206-211, 222).

[22] For a general discussion of wing failures in the Mosquito during high-G manoeuvres, see Brown [42].

tremendous increases in speeds and wing loads that occurred through World War II. If aeroplane construction had followed a non-metallic rather than a metallic path, the technical history of aviation would have been different. But present-day metal aeroplanes can not explain the choice of metal in the past.

Now I arrive at the heart of my argument. Metal succeeded not because the technical case for it was compelling, but rather because advocates of metal portrayed an imagined future that proved compelling within the aeronautical community. Supporters of wood construction, in contrast, completely failed to produce an alternative vision of the future. Supporters of metal achieved their success by exploiting established symbols of technological culture, symbols that linked metal with progress and wood with tradition.[23]

Advocates of metal aeroplanes did not see any ambiguity in the choice of materials. They insisted that metal would eventually prove superior in weight, cost, safety and durability, even if they had little empirical evidence to support their claims. Advocates of metal were not engaged in subterfuge, but rather doing what promoters of new technologies normally do. Support for new technologies must always be based at least in part on future promise rather than demonstrated results. Realistic comparisons between new and established technologies can only occur after the new technology has received a strong push down the learning curve, that is, after it has moved from innovation to diffusion. But this push down the learning curve requires producers and users to make a commitment to the new technology before realistic comparisons are available.

Advocates of metal in the early 1920s understood this dilemma. Like all partisans of particular technologies, they deployed all the rhetorical resources at hand to convince others to provide the material support necessary to make their dreams real. A key part of the pro-metal argument was the mapping of the dichotomy between wood and metal onto the opposition between tradition and modernity. To put it simply, advocates of metal linked wood with tradition and metal with modernity, thus creating the expectation that technological progress would produce the triumph of metal over wood.

The 'modern' proved to be a powerful symbol within the aeronautical community. Since at least the late nineteenth century, the embrace of the new has overwhelmed respect for tradition in the rhetoric of technology. Although engineers have often been politically conservative, they are rarely technologically conservative. Or to be more precise, few engineers are committed to technological conservatism as an ideology, even though many are conservative in practice. In the context of the twentieth-century faith in technological progress, 'tried and true' is weak rhetoric compared with 'new and improved'.

[23] The following discussion is based mainly on Schatzberg [34] (58-63).

Throughout the 1920s, advocates of metal used two main strategies to link wood with tradition and metal with modernity. First they made a historical argument, drawing parallels between the past transitions from wood to metal, especially in transportation. This past trend to metal, they insisted, made the metal aeroplane inevitable. 'All the history of engineering', insisted the prominent British aeroplane engineer John D. North, 'relates the gradual displacement of timber by lighter and more durable structures of steel.' William Stout, an American promoter of metal aeroplanes, and M.E. Dewoitine, a prominent French designer, both invoked the shift from wood to metal ships. In a 1923 article, two American engineers insisted that the aeroplane would follow the shift from wood to steel railway coaches. Steel railway coaches initially weighed more than those of wood, but 'when designers became more experienced and specialized, the steel railway coach became lighter than the wooden coach'. These two engineers were implicitly recognising learning curve effects, and using the principle of the learning curve to argue in favour of metal despite its disadvantage in weight [43, 44, 45, 46].

In all these historical analogies, metal symbolised technological progress, the triumph of the modern over the old. But advocates of metal used a second strategy to link wood with tradition and metal with modernity, arguing that wood represented craft and metal science. Dewoitine, Junkers and many others portrayed wood as an unscientific material, variable, unreliable, imperfectly elastic, and limited to shapes provided by nature. Metal, in contrast, was 'scientific' because of its uniformity, isotropy and elasticity, which provided a better fit to the assumptions used in the stress calculations, see Stout [47], Miller and Seiler [46] (210) and Junkers [48]. In addition, advocates of metal linked wood with craft methods in contrast to the rigorous calculation and planning required by modern industry. According to the French aeroplane designer M.E. Dewoitine, wood was 'a material essentially ideal for the inventor, who .. obtained results with but little design and calculation', whereas metal required the support of a strong engineering department. A spokesman for the US Army was even more explicit. He claimed that 'flying started as an art', but was now 'crying out to science', while 'the finger of science ... pointed to metal', see Dewoitine [45] (5-6) and McDarment [49]. This argument derived its force from the assumption that technological progress involves a shift from traditional craft methods to rigorously scientific procedures. Whether or not this was true, advocates of metal failed to explain why wood structures would not also benefit from scientific investigation.

These arguments proved so powerful because they drew on the established symbolism of industrial culture in the early twentieth century. In political economy, architecture and fine arts, wood was identified with tradition and metal with modernity. Werner Sombart and Lewis Mumford both viewed industrial technology as involving a shift from the organic to the inorganic, from wood to metal. Critics of industrialism like John Ruskin and William Morris praised traditional materials like wood and stone, while condemning new techniques such as the use of cast iron for mass-produced ornamentation. In a mirror image of Ruskin, Le Corbusier and other modernist architects rejected 'heterogeneous and unreliable natural materials' in favour of artificial materials like steel or reinforced concrete [50]. In other words, advocates of metal were able to tap into a broad, pre-existing network of symbolic meanings that linked wood with tradition and metal with industrial progress.

On one level this symbolism did reflect historical reality, but on another level this symbolism was quite ideological, essentialising and distorting the relationship between materials and technological change. The shift from wood to steel did indeed improve many technologies, and make possible structures that would have been impractical in wood. Yet there was nothing in wood that made it unsuited to the machine age, see Schatzberg [34] (53).[24] Beginning in the nineteenth century, wood has been thoroughly industrialised, with machinery and quantity production methods used at all stages of production. Even in the twenty-first century, wood remains an essential structural material in industrialised countries.[25]

Within the aeronautical community, the debate over aircraft materials was framed in terms of wood versus metal. In a technical sense this dichotomy is curious, because aeroplane designers had to choose specific materials, not generic 'wood' or 'metal'. No one proposed fabricating aeroplane structures from cast iron or corkwood. Instead, the choice was between very specific varieties of these materials with good ratios of strength to weight, most importantly aluminium alloys and spruce. The emphasis on wood and metal as general categories provides further evidence of the symbolically driven, ideological character of the debate.

The specific material that triumphed in aeroplanes, aluminium, benefited on its own from symbolic links with modernity. A rising crescendo of voices in the late nineteenth century hailed aluminium as the metal of the future. In 1893 an editorial in a British magazine rhapsodised on the wondrous new metal. Just as 'the world has seen its age of stone, its age of bronze, and its age of iron, so it may before long have embarked on a new and even more prosperous era – the age of aluminium'. Writers praised aluminium for its beauty, lightness, corrosion resistance and abundance. The success of the electrolytic Hall-Héroult process linked aluminium with electricity, another evocative symbol of technological progress. Writers also identified aluminium as a product of modern science, distinctly more modern than other common metals that were discovered in antiquity [54, 55, 56 (quote)]. Advocates of aluminium predicted a new era of lightweight structures, such that 'the Eiffel Tower as a constructive feat would sink into insignificance' [57]. Although aeronautical engineers rarely discussed aluminium in these terms, the aluminium industry no doubt gained considerable strength from the identification of aluminium with technological progress.

---

[24] In fact, the multi-billion dollar industry of 'engineered wood' demolishes the claim that wood is an unscientific material suited only to craft methods. For an overview of this industry, see [51].

[25] Americans, for example, use roughly comparable amounts of wood and steel for 'structural' purposes, broadly construed. For statistics on wood and metals consumption and production by country, see [52, 53].

Overall, this rhetoric established strong connections between metal and modernity in the debate over aircraft materials in the 1920s. But did these symbolic meanings really shape technical choices, or was it just sound and fury? This is a question of historical causation, but unlike laboratory sciences, historians can not isolate causes through experiment. Instead, an argument for historical causation implies a counterfactual analysis, a kind of thought experiment in which one imagines what would have happened if the cause had been absent.

Let us imagine what it was like to be the officer in charge of aeroplane engineering in the US Army in 1920, Major Thurman H. Bane. Bane headed McCook Field, the predecessor to Wright-Patterson Air Force Base in Ohio. Bane supervised over 1000 civilians and military men, including some of the nation's best aeronautical engineers. Bane had a large but declining budget for research and development, and he had to make tough choices about how to allocate this budget.

What evidence could advocates of metal have used to convince a tough-minded engineering officer like Bane to devote substantial resources to metal aeroplanes? Could these advocates have demonstrated from first principles or empirical evidence that metal aeroplanes would weigh less than wood aeroplanes? Could they have substantiated claims that metal aeroplanes would cost less to produce than wood? Could they have used known material properties, like fatigue strength and burn-through rates, to show that metal aeroplanes would be safer than wood? Could they cite even preliminary field tests suggesting that metal aeroplanes had lower maintenance costs than wood? Could they show that wartime supplies of metal were likely to be more secure than supplies of wood?

Advocates of metal could have answered none of these questions in the affirmative, see Schatzberg [34] (67-69). Yet advocates of metal did not merely argue for modest research funding in order to acquire the evidence needed for technically informed answers to these questions. Instead, advocates of metal invoked the rhetoric of technological progress, describing the Junkers all-metal JL-6 as the 'airplane of the future', and demanding that the Air Service make an immediate full-scale commitment to develop metal aeroplanes. These arguments persuaded a joint Army-Navy technical committee to endorse the immediate 'acquisition and construction of all-metal airplanes ... by both the War and Navy Departments'. In August 1920 Major Bane produced a new budget proposal that devoted roughly half of the airframe R & D funds to metal construction. At the same time, Bane immediately withdrew support for wooden aeroplane research at the Forest Products Laboratory, arguing that such research would become irrelevant with the shift to metal construction, see Schatzberg [34] (41, 68, 128).

This specific moment nicely captures the circular, self-fulfilling nature of support for metal construction, and the powerful role that expectations play in shaping technical choice. In August 1920, Bane had little evidence for even the potential superiority of metal aeroplanes. Yet the symbolic connection between metal and modernity convinced Bane that the shift to metal was inevitable. Based on this expectation,

Bane decided to move R & D funds from wood to metal construction, thus making it significantly more likely that metal would surpass wood as a material for aircraft construction.

Such expectation-driven shifts in R & D efforts were repeated throughout the American aeronautical community, within government agencies, among manufacturers, and in universities. Despite repeated and expensive failures, the US Army and Navy continued to fund metal aeroplane projects at far higher levels than wood projects, both for developing new aeroplanes and for research into problems of design and construction. The US National Advisory Committee for Aeronautics focused intensively on problems related to metal, such as inter-crystalline corrosion, while neglecting comparable problems in wood structures, such as the durability of glues. American manufacturers devoted considerable technical resources to improving the design and production of metal aeroplanes, often suffering huge losses as a result, most notably in the case of Henry Ford. From about 1930 on, aeronautical engineers conducted extensive empirical research to optimise the design of reinforced aluminium alloy shell structures, while largely ignoring comparable issues in plywood stressed-skin designs. The tremendously successful Douglas DC-1, 2 and 3 series, which aviation historians commonly view as the turning point in all-metal construction, was really the end point of a roughly 15-year commitment to the metal aeroplane. Without this multifaceted expectation-driven R & D effort, it is unlikely that metal structures would have been able to dominate high-performance aeroplanes by World War II, see Schatzberg [34] (chaps. 4-6).

The vast engineering resources devoted to the metallic path resulted in the aluminium alloy, stressed-skin monoplanes structures that helped make aviation a key technology of modern civilisation. Yet the success of metal was not without cost. By focusing efforts on perfecting metal structures, the aeronautical community failed to explore potentially fruitful developments in non-metallic materials. And by seizing the rhetoric of progress for themselves, advocates of metal made it much harder for promising non-metallic materials to obtain R & D support.

Non-metallic materials: a neglected path

At the beginning of this paper I suggested that the history could only provide indirect aid to people developing new technologies by contributing to the practical wisdom required for good technical judgements. The shift from wood to metal aeroplanes is relevant in just this way to present-day debates over composite materials, including hybrid materials like Glare. The history of composite materials has only begun to be written, so my remarks must remain provisional. Although Glare emerged primarily from research on aircraft metals, it has affinities with two related traditions in the history of non-metallic aeroplane materials, i.e. fibre-reinforced plastics (FRPs) and sandwich structures. Both traditions emerged between the world wars, but neither research path received much attention before World War II.

The first tradition, fibre-reinforced plastics for aeroplane structures, was closely connected with attempts to develop improved aeroplane woods. Beginning in the late 1920s, researchers at the *Deutsche Versuchsanstalt für Luftfahrt* (DVL) began studying commercially available thermosetting resins reinforced with cellulose fibres in various forms, such as sawdust or cotton cloth. The DVL researchers quickly discovered that these materials suffered from low specific stiffness, so they shifted to using very thin wood veneers to take advantage of wood's higher stiffness, laminating these veneers with varying amounts of phenolic resin. By the mid-1930s this small research project had produced a material with quite promising technical properties compared to aluminium alloys. More significantly, however, the DVL research showed how blurred the boundary was between resin-bonded plywood and fibre-reinforced plastics. Whether the resin was reinforced with powdered wood or thin veneers did not seem to require a shift in categories between 'wood' and 'plastic', see Schatzberg [34] (179-181). The de Havilland Aircraft Company conducted similar research from the mid-1930s; de Havilland happened to be one of the few remaining British manufacturers of high-performance wood aircraft [58].[26]

By the late 1930s, both the promise and problems of fibre-reinforced plastics were clear, at least to well-informed researchers. The promise lay in the high specific tensile strength of common fibres like cotton and silk, several times higher than that of aircraft metals. By combining these fibres with a synthetic resin matrix, researchers hoped to produce materials with specific strength properties comparable to aircraft metals at significantly lower density. Researchers quickly found that they could improve the specific strength of FRPs to match those of aluminium alloys in at least one direction. But even when strength properties were promising, these materials proved substantially less stiff (E/sg) than wood or metal, especially in compression. Even before World War II, increasing the stiffness of FRPs had become one of chief goals of plastics researchers.[27]

The aeronautical community expressed surprisingly little interest in FRPs during the 1930s, despite clear indications of potential promise. One reason for this lack of interest was the symbolic link between FRPs and wood. Researchers at the DVL and de Havilland both discovered that the most promising 'plastics' were in fact laminations of very thin wood veneers with thermosetting resins. In the United States, this line of materials research was taken up by a small aeroplane company, Fairchild, and an innovative plywood manufacturer, Haskelite. Together, these two companies developed a system of moulded resin-bonded veneers marketed as 'Duramold'. In 1937 this material was used to make the fuselage of a five-place commercial aeroplane, the Fairchild F-46. Duramold was promoted as a radical new material, but the chief materials scientist for Army aviation insisted that Duramold was merely plywood with a new adhesive. Despite the promise of substantial manufacturing efficiencies with the Duramold system, the US Army decided not to

[26] See also the comments by de Havilland engineers E.P. King and C.C. Walker in De Bruyne [59]. During World War I, Caldwell developed 'Micarta', a unidirectional fibre-reinforced Bakelite material for propellers, but this material was apparently never considered for use in airframes [60].

[27] A good summary is provided by Kline [61].

fund further development of Duramold but instead to 'concentrate on the perfection of metal airplanes' [62]. Without military support, neither Fairchild nor Haskelite had the resources necessary to develop Duramold further, see Schatzberg [34] (181-187).

The materials shortages of World War II revived research in FRPs. This research was at first not focused on fundamental advances, but rather on quickly developing materials that could serve as direct substitutes for scarce aluminium alloys. In Britain the Ministry of Aircraft Production established a plastics committee that had jurisdiction over all non-metallic materials, including wood, thus illustrating the continuing link between wood and plastics; in the United States the NACA Committee on Miscellaneous Materials had a similar broad mandate.[28] By mid-1943, however, the aluminium shortage had eased considerably, and research shifted to the development of fundamentally improved materials. The key technical shift was from cellulose to inorganic fibres, asbestos in Britain and glass fibres in the United States. Glass fibres used polyester resins as the matrix, which did not need the high temperatures and pressures required for phenolic resins, see Schatzberg [34] (226).

Despite some promising results with glass fibre plastics, there was little room for such materials in the postwar era of jet engines, rockets, and supersonic flight. Even though fibreglass performed relatively well at elevated temperatures, the material did not improve on the low elastic modulus of cellulose fibre plastics, which made fibreglass unsuitable for high-performance aircraft. By the 1950s it became clear that fundamental improvements in fibre-reinforced plastics depended on using fibres of much higher stiffness than traditional materials, see Schatzberg [34] (226-227) and Hoff [65] (52).

The Cold War provided the context that made this research possible. During the Korean War, military R & D spending in the United States rose above World War II levels in real terms, and remained at these high levels into the 1960s. Although the cutting edge of materials research lay in high-temperature applications for supersonic flight and re-entry vehicles, significant funding was also available for long-range development of new materials in the emerging field of composites.

Fundamental research soon identified several brittle solids with specific moduli many times greater than traditional materials. The problem was how to produce the thin filaments or whiskers required to develop the strength of these often exotic materials, some of which were highly toxic, like beryllium, while all were tremendously expensive. In the United States, the Materials Laboratory at Wright Field focused on boron fibres, which were produced by vapour deposition at high temperatures. By the mid-1960s, boron fibres cost $600/pound ($1320/kg), with little prospect for major price reductions. Boron fibre composites provide a clear example of what Mary Kaldor has called 'baroque' technologies that could only be developed by well-funded military researchers isolated from the civilian market. The

[28] For Britain, see [63]. For the US, see for example [64].

British, in contrast, focused on a more mundane substance, carbon, which was also one of the most promising in terms of specific stiffness. Scientists at the Royal Aircraft Establishment began working on carbon fibres in the early 1950s, developing a commercially viable production process by 1964. In 1965, an US Air Force study predicted that these new high-stiffness composites would transform aeroplane structures and bring about 35 percent weight savings compared to metal, see Hoff [65] (53), McMullen [66] and Schatzberg [34] (227-228, 230).

Despite the apparent promise of carbon fibre composites, they have failed to displace aluminium alloys in the principal structures of commercial aeroplanes. Glare, however, does not rely on high-stiffness fibres for its advantages. Instead, Glare combines dissimilar materials in a way that has conceptual affinities with a second tradition in non-metallic aircraft materials – sandwich construction. The principle of sandwich construction is quite simple, to combine a low-density core with high-density faces in order to increase the stability of shell structures. In this way, the high-density material is placed farther from the neutral axis where it can carry more stress in bending, while being stabilised against buckling by the low-density core. Sandwich structures represent an attempt to capture the buckling advantages of low-density materials while retaining the strength and stiffness of high-density materials [67].

Perhaps the most famous aviation application of sandwich structures was the de Havilland Mosquito, one of the most formidable warplanes of World War II and a triumph of wood engineering. The fuselage skin was a built-up structure consisting of thin birch plywood over a balsa wood core. The stability of this thick, stiff skin allowed designers to dispense with longitudinal stiffeners, see Schatzberg [34] (214-215). The Mosquito's success stimulated research in other sandwich materials. The first American application of glass-fibre-reinforced plastics to aeroplane structures was in a sandwich structure that consisted of glass fibre face layers over a balsa wood core. In 1943, engineers at Wright Field used this material to build a monocoque fuselage for the Vultee BT-15 trainer. After the war, research shifted to finding suitable synthetic alternatives to balsa for the core material [68]. Burt Rutan's *Voyager* aeroplane, see Figure 4, which completed its non-refuel non-stop round-the-world flight in 1987, is heir to this research. The fuselage shell of the *Voyager* consists of a honeycomb core made of Nomex paper covered with thin sheets of carbon fibre composites [69].

Sandwich structures offered great promise for non-metallic materials, but nothing prevented the use of metal in the face plies. Such an approach was pursued in the early 1920s by the Haskelite Manufacturing Corporation of Chicago, a major supplier of aircraft-grade plywood. Haskelite developed an aluminium-faced plywood sheet, which it marketed under the name 'Plymetl'. Haskelite claimed that Plymetl was 50 to 100 times more resistant to buckling than sheet metal of the same weight. Despite favourable publicity in the trade press, there is no evidence that Plymetl was ever used in aeroplane structures. Later in the decade Goodyear developed a similar metal-faced sandwich material with an expanded rubber core instead of plywood, but it was apparently never commercialised [70, 71, 72, 73]. During World War II Chance-Vought did some research on aluminium/balsa sandwich structures and built at least one aeroplane, the XF5U-1, using this material [74].

These metal-faced sandwich structures never found widespread use in aircraft structures. The failure of these innovations to meet with commercial success is not surprising. A high conceptual boundary separates metallic and non-metallic materials. The two categories have fundamentally different material properties, require distinct skills for manufacturing and repair, and carry incompatible symbolic meanings. Although Glare is considered a laminate rather than a sandwich, since it lacks a low-density core, it shares a similar conceptual boldness with these earlier attempts to combine the advantages of metallic and non-metallic materials.[29]

Figure 4: Voyager *round-the-world aeroplane* [75].

Despite predictions of a composites revolution going back some thirty years, non-metallic materials have made only modest inroads in commercial aviation, especially for large airliners.[30] Why have these potentially advantageous non-metallic materials failed to find a larger place in commercial aviation, either in combination with metal or on their own?

This question seems particularly puzzling in technical terms. Sandwich structures promised substantial economy in production costs by eliminating the need for local skin stiffeners. Composites promised huge weight savings as well as significant manufacturing efficiencies, see Yaffee [77] (38).[31] Compared with advocates of metal in the 1920s, proponents of composites in the 1960s were able to make a much stronger case for the potential benefits of the new materials. It took no more than 15

[29] Thanks to Ad Vlot for clarifying for me the difference between laminates and sandwiches.
[30] See for example Von Braun [76].
[31] See also Hoff [78] (12).

years for metal to replace wood as the dominant aeroplane material, but over 35 years of composites development have barely dented the dominance of metal, at least in large commercial aircraft.

There are good reasons for the slow growth of composites. One is simply the distance that metal aeroplanes have travelled down the learning curve, resulting in a body of knowledge quite specific to light-alloy reinforced-shell structures. Since the early 1930s, aeronautical engineers have focused on the design of these metal structures, producing a massive body of empirical data and a wealth of analytical tools. Aircraft manufacturers have improved aircraft production to a fine art, despite the continued high labour costs of assembling riveted reinforced shell structures. Airlines have developed extensive maintenance systems that can keep fatigue-prone metal aeroplanes operating safely for decades. Government regulations ensure that new metal aeroplanes are designed with adequate strength.

A shift to a radically new material would disrupt every one of these well-developed systems. Composite materials require new design tools as well as new structural forms. Manufacturing methods are radically different, as are procedures for inspection and repair. Existing government regulations may not be adequate for assessing the safety of the new design and manufacturing techniques. In other words, adopting composite materials would make thousands of engineer-years of accumulated experience obsolete. When advocates of metal sought to displace wood in the 1920s, none of these knowledge systems existed in more than an incipient form.

The growth of composites has also been limited by the aggressive response of aluminium firms. Until the 1980s, there had been no major improvements in the wrought aluminium alloys used in most aircraft structures since World War II. The widely-used 2024 alloy, for example, was developed by Alcoa in the 1930s. Yet the potential threat from composites spurred the aluminium industry to invest hundreds of millions of dollars in developing new esoteric alloys like aluminium-lithium, which significantly reduce the weight advantages of composites [79].

The late aeronautical engineer Nicholas Hoff has suggested a third impediment to the adoption of new non-metallic composites – American product liability laws. Hoff claimed that the doctrine of strict liability made aircraft manufacturers reluctant to employ new materials, since unforeseen problems almost invariably emerge in actual airline operations. According to Hoff, American aeroplane manufacturers in the 1930s were willing to embrace metal despite ‘woefully inadequate’ design information, especially regarding the buckling behaviour of reinforced shells, see Hoff [78] (12-13).

Changes in American product liability laws hardly seem a sufficient explanation, however. Since the late 1920s, the aviation industry has recognised that commercial aviation can not succeed unless the public remains confident in the safety of flying. The airlines and large manufacturers in essence invited the US government to regulate flight safety, producing the Air Commerce Act of 1926. Since then, design-related structural failures have been very rare in aeroplanes certified by the FAA or similar agencies in other countries [80, 81].

There is, nevertheless, one final difference that may help explain the slow adoption of composites compared to metal. Supporters of composites never succeeded in creating a sense of inevitability for their materials, never managed to make composites seem like a moral necessity in the onward march of technological progress. Although the field of composites includes all mechanical combinations of dissimilar materials, including ceramics and metals, in practice aircraft composites mean fibre-reinforced plastics. However impressive the material properties of carbon fibre materials, plastics carry an ambiguous cultural legacy. Since the 1930s, the plastics industry has self-consciously sought to build a symbolic link between plastics and technological progress, promoting an image of plastics as aesthetically modern. Yet especially since World War II, the public perception of plastics as cheap substitutes has remained strong. In the 1960s, 'plastic' became a countercultural synonym for unauthenticity, see Schatzberg [34] (230).[32]

The ambiguous cultural legacy of plastics almost certainly helps explain the popularity of the term 'composites' among advocates of advanced fibre-reinforced plastics since the mid-1960s. But not even Wernher von Braun's endorsement could give composites the same cultural urgency as metal [76].[33] In the opposition between metal and wood, metal had the advantage of competing against a material marked culturally as pre-industrial. The wooden aeroplane, in fact, never really made sense culturally, symbolising at the same time the modernity of flight and the traditionalism of wood. Since the Douglas DC-3, 'metal' and 'aeroplane' have become linked in symbolic culture. Although the high-tech aura of aluminium has faded somewhat with the metal's ubiquitous presence in mundane consumer goods, aluminium and aerospace remain symbolically linked. Despite their technical promise, composites have no symbolic advantage over metal comparable to metal's symbolic advantage over wood. Without this advantage, composites face a catch-22, having to prove their superiority in practice before being widely adopted, but needing to be widely adopted in order to develop superiority in practice.

[32] By far the best discussion of cultural attitudes towards plastics in the United States is Meikle [82]. Attitudes in Europe may have been different, however.

[33] British plastics researchers during World War II were already using the term 'composites' in a general sense. In the interwar period, the term 'composite' was commonly used for mixed wood and metal aeroplanes, typically those with wooden wings and steel-tube fuselages. During World War II, British plastics researchers sometimes referred to FRPs and resin-bonded veneers as 'composite' materials. See for example the reference to 'high-density composite plastic veneer materials' in E. Reeve Angel to G.K. Dickerman [83]. The term began to be used in its broader sense by the late 1950s, see for example [84]. Not until the mid-1960s, however, did the term become widely identified with all varieties of fibre/matrix materials, including FRPs, see for example Yaffee [77] (38-48+).

It is in this light that one can appreciate Glare's symbolic as well as material advantages.[34] With no symbolic wave to ride into the future, new aeroplane materials are more likely to succeed by accommodating themselves to the existing practices and cultural meanings of the aeroplane. Glare meets this challenge by taking a middle path between metals and composites, creating both a symbolic and material compromise suited to commercial aviation. In many ways Glare is an ideal post-Cold War material, being focused primarily on reducing the operating costs of commercial aircraft. Military requirements have driven the development of advanced composites, pushing the boundaries of strength and stiffness while demanding a complete transformation of aircraft design, construction and maintenance. Glare, in contrast, gains far more in familiarity and ease of use than it loses by not pushing the performance envelope. Glare combines materials which have been used for decades, that is, standard aluminium alloys and fibreglass, for which there is a wealth of information based on aircraft applications. Glare's clearest advantage, fatigue strength, makes it ideally suited to the decades of intensive use required by commercial aircraft.

The role of symbolic meanings in technological change has both positive and negative implications for the success of Glare. On the one hand, Glare does not benefit from strong associations with a technological wave of the future. There is no Glare.com; Glare will not help build the Internet and does not rely on biotechnology. Instead, Glare uses established materials in a conceptually bold combination to achieve a significant improvement in a mature technology, the commercial airliner. Yet Glare's hybrid nature gives it a symbolic flexibility that its supporters should not hesitate to employ. Glare can be represented as a revolutionary development that brings out the best qualities of both metals and composites. But Glare can also be represented as an incremental improvement over existing aircraft materials, one designed to cause minimal disruptions to existing methods of aircraft design, construction and operation. These contrasting symbolic meanings are both aspects of the truth, but they provide distinct advantages in different contexts. Will these meanings help or hinder Glare's wider use? Historians should not try to answer such questions.

References

[1] Hegel, *Grundlinien der Philosophie des Rechts*, 1821, quoted in: 'Eule der Minerva', in: *Historisch Kritisches Wörterbuchs des Marxismus*, http://www.argument.de/wissenschaft/eule.html.

[2] Robert Graves, *Greek Myths*, rev. ed., New York: Penguin Books, 1960, 1: 96.

[3] Bruno Latour, *Science in Action: How to Follow Scientists and Engineering through Society*, Cambridge: Harvard University Press, 1987.

[34] My analysis of Glare's symbolic significance is somewhat speculative, since I lack temporal perspective as well as access to internal documents, both essential to the historian's task.

[4] Georg Wilhelm Friedrich Hegel, *Hegel's Philosophy of Right*, trans. T.M. Knox, Oxford: Clarendon Press, 1952, preface.

[5] Richard J. Bernstein, *Beyond Objectivism and Relativism: Science, Hermeneutics, and Praxis*, Philadelphia: University of Pennsylvania Press, 1983.

[6] David Pye, *The Nature and Aesthetics of Design*, London: Herbert Press, 1978.

[7] Michael Polanyi, 'Skills', chap.4 in: *Personal Knowledge: Towards a Post-Critical Philosophy*, Chicago: University of Chicago Press, 1958: 49-65.

[8] *Studies in History and Philosophy of Science*, 30, 1999: 137-171.

[9] John Staudenmaier, 'Recent Trends in the History of Technology', in: *American Historical Review,* 95, 1990: 717, 723.

[10] Trevor J. Pinch and Wiebe E. Bijker, 'The Social Construction of Facts and Artifacts: Or How the Sociology of Science and the Sociology of Technology Might Benefit Each Other', in: *Social Studies of Science,* 14, 1984: 399-441.

[11] W. Brian Arthur, 'Positive Feedbacks in the Economy', in: *Scientific American*, 262, Feb. 1990: 92-99.

[12] W. Brian Arthur, 'Increasing Returns and the New World of Business', in: *Harvard Business Review*, 74, July-Aug. 1996: 100-109.

[13] W. Brian Arthur, 'Competing Technologies, Increasing Returns, and Lock-in by Historical Events', in: *Economic Journal*, 99, 1989: 116-131.

[14] Karl F. Habermeier, 'Competing Technologies, the Learning Curve, and Rational Expectations', in: *European Economic Review*, 33, 1989: 1293-1311.

[15] T.P. Wright, 'Factors Affecting the Cost of Airplanes', in: *Journal of the Aeronautical Sciences*, 3, February 1936: 124.

[16] Paul A. David, 'Understanding the Economics of QWERTY: The Necessity of History', in: *Economic History and the Modern Economist*, ed. William N. Parker, Oxford: Basil Blackwell, 1986: 30-49.

[17] S.J. Liebowitz and Stephen E. Margolis, 'Fable of the Keys' in: *Journal of Law and Economics*, 33, 1990: 1-26.

[18] John Cassidy, 'The Force of an Idea', in: *New Yorker*, 73, Jan. 12, 1998: 35.

[19] Charles Bazerman, *The Languages of Edison's Light*, Cambridge: MIT Press, 1999.

[20] Richard Rhodes, *The Making of the Atomic Bomb*, New York: Simon & Schuster, 1986.

[21] Richard E. Rowberg, *Congress and the Fusion Energy Sciences Program: A Historical Analysis*, CRS Report RL30417, Washington DC: Congressional

Research Service, 2000 [available on-line at http://www.pppl.gov/common_pics/rowberg_rpt.pdf].

[22] Donald MacKenzie, 'From the Luminiferous Ether to the Boeing 757: A History of the Laser Gyroscope', in: *Technology and Culture*, 34, 1993: 514-515.

[23] Photo courtesy Douglas Products Div., Boeing Company, Long Beach CA, photo SM5807.

[24] W.F. Durand, 'Some Outstanding Problems in Aeronautics', in: US National Advisory Committee for Aeronautics, *Fourth Annual Report*, Washington: GPO, 1920.

[25] J.E. Gordon, 'The Design of Materials', in: Proceedings of the Royal Society of London, Series A, Mathematical and Physical Sciences, 319, 1970: 137-143.

[26] J.E. Gordon, 'Some Considerations in the Design of Engineering Materials Based on Brittle Solids', in: Proceedings of the Royal Society of London, Series A, Mathematical and Physical Sciences, 282, 1964: 16-23.

[27] J.E. Gordon, *The New Science of Strong Materials, or Why You Don't Fall through the Floor*, 2nd ed., Princeton NJ: Princeton University Press, 1976.

[28] Edward P. Warner, discussion comment to 'Symposium on Metal Aircraft Construction', in: *SAE Journal*, 22, April 1928: 433.

[29] T.P. Wright, 'Relation between Commercial Airplane Design and Commercial Uses of Airplanes', in: *ASME Transactions, Aeronautical Engineering*, 51, 1929: 201.

[30] Fortant, 'The Preliminary Design of Aircraft', in: *Aviation*, 13, Sep. 25, 1922: 381-382 [trans. of 'L'Aérotechnique et le problème de la conception des aéronefs', in: *L'Aéronautique*, Nov. 15, 1921].

[31] Ivo Edwards and F. Tymms, *Commercial Air Transport*, London: Isaac Pitman & Sons, 1926, 38-39.

[32] Alfred A. Gassner, 'Weight Saving by Structural Efficiency', in: *SAE Journal*, 27, Oct. 1930: 466.

[33] Karl Thalau, 'Aufgaben der Luftfahrzeug-Statik', in: *Deutsche Versuchsanstalt für Luftfahrt, Jahrbuch*, 1931: 67.

[34] Eric M. Schatzberg, *Wings of Wood, Wings of Metal: Culture and Technical Choice in American Airplane Materials, 1914-1945*, Princeton NJ: Princeton University Press, 1999, 17-18.

[35] John H. Morrow Jr, *German Air Power in World War I*, Lincoln: University of Nebraska Press, 1982, 190-191.

[36] Eric M. Schatzberg, 'The Airplane in the University: Industry-University Relations in Aeronautical Engineering on the West Coast between the World

Wars', in: *Proceedings of the 1988 Annual Conference of the American Society for Engineering Education*, ed. Lawrence P. Grayson and Joseph M. Biedenbach, Washington DC: ASEE, 1988, 4: 1705-1715.

[37] Charles D. Bright, 'Machine Tools and the Aircraft Industry: The Boeing Case', in: *Journal of the West*, 30, 1991: 50-57.

[38] *Air Corps Technical Report*, no. 2895, Oct. 1, 1927, Library of Congress.

[39] A. Vlot, L.B. Vogelesang and T.J. de Vries, 'Towards Application of Fibre Metal Laminates in Large Aircraft', in: *Aircraft Engineering and Aerospace Technology*, 71, no. 6, 1999: 558-570.

[40] Joseph S. Newell, comment to H.V. Thaden, 'Metallising the Airplane', in: *ASME Transactions, Aeronautic*, 52, 1930: 171.

[41] *US NACA Technical Note*, no. 283, April 1928.

[42] Vernon Brown, 'Mosquito Structural Failures', Service Accident Report No. Misc. 23E, Jun. 7, 1944, in: National Archives Canada, RG24, vol. 5406, file HQS-60-11-36, *Mosquito Aircraft, Reports on Structural Failures*.

[43] John D. North, 'The Case for Metal Construction', in: *Journal of the Royal Aeronautical Society*, 28, 1923: 3-25.

[44] William B. Stout, 'The Modern Airplane and All-Metal Construction', in: *SAE Journal*, 11, Dec. 1922: 495-504.

[45] M.E. Dewoitine, 'The Metal Construction of Airplanes - Its Advantages - Its Present State - Its Future', in: *US NACA Technical Memorandum*, No. 349, Feb. 1926.

[46] Roy G. Miller and F.E. Seiler, 'The Design of Metal Airplanes', in: *Aviation*, 14, Feb. 19, 1923: 210-214.

[47] William B. Stout, 'Wood Versus Metal for Airplanes', in: *US Air Service*, 8, May 1923: 16.

[48] Hugo Junkers, 'Metal Aeroplane Construction', in: *Journal of the Royal Aeronautical Society*, 28, Sep. 1923: 417.

[49] Corley McDarment, 'Will the Future Airplane Be of Metal?', in: *Iron Age*, 115, Jan. 1, 1925: 20-21.

[50] Le Corbusier-Saugnier, 'L'esthétique de l'ingénieur: maisons en série', in: *Nouvelle Esprit*, no. 13, Dec. 1921: 1530.

[51] http://www.apawood.com, website of the Engineered Wood Association.

[52] US Geological Survey, *Minerals Yearbook*.

[53] United Nations Food and Agriculture Organisation, *FAO Yearbook*.

[54] Helmutt Maier, 'New Age Metal or Ersatz? Technological Uncertainties and Ideological Implications of Aluminium up to the 1930s', in: *ICON*, 3, 1997: 181-201.

[55] Marie Boas Hall, 'The Strange Case of Aluminium', in: *History of Technology*, 1, 1976: 145, 148.

[56] 'The Future of Aluminium', in: *The Spectator*, 71, July 1893: 76-77.

[57] Amos W. Wright, 'The Age of Aluminum', in: *Harper's Weekly*, 34, 1890: 707-708.

[58] G. de Havilland, ''Filled' Resins and Aircraft Construction', in: *Journal of the Aeronautical Sciences*, 3, 1936: 356-357.

[59] N.A. de Bruyne, 'Plastic Materials for Aircraft Construction', in: *Journal of the Royal Aeronautical Society*, 41, 1937: 583-589.

[60] 'Bakelite Micarta Propellers', in: *Aviation*, 8, 1920: 354-357.

[61] Gordon M. Kline, 'Plastics as Structural Materials for Aircraft', in: *Journal of the Aeronautical Sciences*, 5, 1938: 391-396.

[62] Brig.Gen. George R. Spalding (Asst. Chief of Staff) to Chief of Staff, 'Plastic Materials', Feb. 17, 1938, General Correspondence, 1917-38, Office of the Chief of the Air Corps, 1917-44, Records of the Army Air Forces, Record Group 18, National Archives, Washington DC, box 939, 423A-Plastic Materials.

[63] 'Plastics Sub-Committee, Aeronautical Research Committee, Minutes (DRAFT)', Jan. 20, 1940, DSIR, 22/116, Records of the Department of Scientific and Industrial Research, Public Record Office, Kew, Surrey, England.

[64] 'Minutes of Meeting of Subcommittee on Miscellaneous Materials and Accessories, Committee on Aircraft Materials', Sep. 24, 1940, National Archives, RG255, NACA Numeric File, Box 229, file 42-13B.

[65] N.J. Hoff, 'Composite Materials in Aircraft Structures', in: *Progress in Science and Engineering of Composites*, ed. Tsuyoshi Hayashi, Kozo Kawata and Sokichi Umekawa, Tokyo: Japan Society for Composite Materials, 1982.

[66] P. McMullen, 'Fibre/Resin Composites for Aircraft Primary Structures: A Short History, 1936-1984', in: *Composites*, 15, 1984: 222-230.

[67] Robert McLarren and Irving Stone, 'Sandwich Structures for Aircraft', in: *Aviation Week*, 47, Nov. 3, 1947: 28.

[68] Charles A. Breskin, 'Plastics Sandwiches', in: *Scientific American*, 173, Sep. 1945: 155-158.

[69] Barton Burnstein, 'The Last Plum', in: *New Yorker*, 62, Aug. 4, 1986: 40.

[70] 'New Aircraft Material', in: *Aviation*, 10, May 2, 1921: 551.

[71] 'Trade Note: Plymetl – A New Engineering Material', in: *Aviation*, 10, May 16, 1921: 625.

[72] *Aircraft Year Book 1925*, New York: Aeronautical Chamber of Commerce of America, 1925, 209.

[73] B.F. Goodrich Co., 'Aeroboard, Dural Faced', April 30, 1928, National Archives, RG255, NACA General Correspondence 1915-42, box 218, file 42-6C.

[74] Eugene E. Lundquist to Engineer-in-Charge, LMAL, Feb. 9, 1944, Re: 'Visit to Chance-Vought Company', National Archives, RG255, NACA General Correspondence, Decimal File 1929-1952, Box 154, file 453.4-Plywood, 1941-1950.

[75] Photo courtesy Dick Rutan, *Voyager* pilot.

[76] Wernher von Braun, 'Lighter Than Aluminum ... Stronger Than Steel!', in: *Popular Science*, 194, Feb. 1969: 98-101+.

[77] Michael L. Yaffee, 'Composite Materials Offer Vast Potential', in: *Aviation Week & Space Technology*, 82, May 3, 1965.

[78] N.J. Hoff, 'Innovation in Aircraft Structures - Fifty Years Ago and Today', in: *A Collection of Technical Papers - AIAA/ASME/ASCE/AHS Structures, Structural Dynamics, and Materials Conference, 25th*, May 14-16, 1984, paper AIAA-48-0840, 12.

[79] Margaret B.W. Graham and Bettye H. Pruitt, *R & D for Industry: A Century of Technical Innovation at Alcoa*, Cambridge: Cambridge University Press, 1990, chap. 10.

[80] Thomas W. Walterman, *Airpower and Private Enterprise: Federal-Industrial Relations in the Aeronautics Field, 1918-1926*, Ph.D. diss., Washington University, 1970, 506-526.

[81] Nick A. Komons, *Bonfires to Beacons: Federal Civil Aviation Policy under the Air Commerce Act, 1926-1938*, Washington: GPO, 1978.

[82] Jeffrey L. Meikle, *American Plastic: A Cultural History*, New Brunswick NJ: Rutgers University Press, 1995.

[83] E. Reeve Angel to G.K. Dickerman, Feb. 5, 1943, AVIA, 15/3256, Records of the Ministry of Aircraft Production, Public Record Office, Kew, Surrey, England.

[84] *Composite Materials and Composite Structures*, Sagamore Army Materials Research Conference Proceedings, 6th, Syracuse NY: Syracuse University Press, 1959.

Response 1

Some considerations for new materials integration into aircraft systems

Flake C. Campbell

Boeing Military Aircraft & Missile Systems

(Disclaimer: The viewpoints offered in this paper are the author's and do not necessarily represent the views of The Boeing Company; however, they do represent 32 years of experience in the aerospace business)

'The first time you hear about a new material is the best time'

I do not remember exactly who told me that about 20 years ago, but after 32 years in the military aircraft business, I have come to believe it. What it simply means is that no single material, no matter how well it is marketed by the eager material supplier, is without faults. The more you work with any new material, the more you find out about its shortfalls.

For example, aluminium alloys have traditionally been a major material used on both commercial and military aircraft. However, aluminium is so prone to corrosion problems that the science of corrosion protection systems has become an engineering field of its own. In addition, the fatigue strength of aluminium is quite frankly not very good when compared to other high-strength materials, such as steel or titanium. Well, since I mentioned steel and titanium, why not use them instead of aluminium. The answer, of course, is that they are used in selective areas, but they are also quite a bit heavier than aluminium. As our keynote speaker said, when it comes to aircraft materials, 'weight is king'. High-strength steels, used primarily in landing gear applications, have their own set of corrosion problems, e.g. hydrogen embrittlement, and have been prone to some quite dramatic brittle failures. Titanium, on the other hand, is naturally quite corrosion-resistant and has better fatigue strength than aluminium. But again, it is heavier than aluminium and much more expensive to form and machine. Well, since composites are lighter than aluminium, do not corrode, and have excellent fatigue properties, why not build the entire airframe out of composites? Unfortunately, composites have their limitations also, e.g. they are expensive, difficult to fabricate and assemble, and have poor resistance to out-of-plane loading, i.e. delamination resistance. Composites like stiff and well-defined load paths.

Well, what about Glare? At this point it is so new that I venture to say we really do not know what its limitations are, but we will painfully discover them along the way. In 20 years or so, we may be using nano-composites. Whatever form they take, they too will have shortcomings. Our grandchildren, or great-grandchildren, may someday develop 'unobtainum', i.e. the perfect material, i.e. lightweight, easy to fabricate and assemble, inexpensive and with infinite strength and stiffness. This will no doubt be followed by the 'next-generation' or improved 'unobtainum', when the problems with the original 'unobtainum' surface. I am certainly not trying to be critical of any of the structural materials we currently use. The materials developed during the 20th century have played a significant role in making the aircraft industry what it is today. I am simply pointing out that no material is perfect or that it can be used for all applications.

Key material selection criteria

In my opinion, there are at least four key elements to selecting a new material for an aircraft application, namely:

1. stable material and material supplier
2. materials and design database
3. stable process
4. demonstrated technology

A *stable material* is one in which the material supplier has finished their development work and has frozen the formulation (chemistry) and has a standard documented procedure for making it, along with the necessary in-process controls in place. By a *stable material supplier*, I mean one that is financially healthy, is large enough to withstand fluctuations in the market, and has committed to this material product form by investing in the technology and production capacity. If any of these elements are missing, you have the potential problem of putting the material into service only to find out later that the supplier can not support your capacity or technology needs.

The second element, *materials and design database*, covers a lot of territory and can be quite an investment. Nevertheless, it is critical to have material allowables, design allowables, understand the influence of the environment, know the effects of defects, and be able to detect them with non-destructive inspection (NDI). This is the one area of the new materials development process that is quite closely monitored, or even mandated, by the regulatory agencies for both commercial and military aircraft.

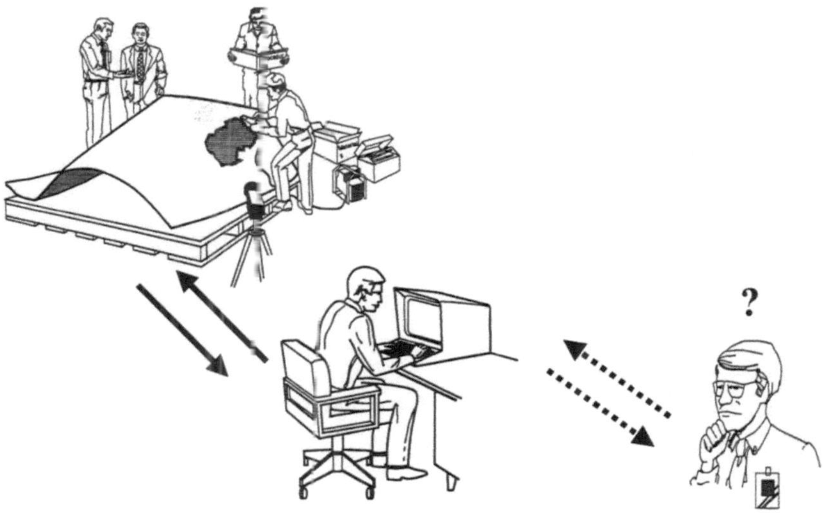

Figure 1: Traditional materials development.

By a *stable process*, I mean that the manufacturing R & D has been thoroughly carried out to understand and be able to control the fabrication and assembly processes once they are put into production. This can also be an expensive investment and one that is usually under-funded, unfortunately. However, fixing a process while you are trying to make rate on aircraft deliveries can prove to be a very expensive and painful proposition.

I also believe that *technology demonstrations* are important. Demonstrating new materials, processes and manufacturing technologies by building sub-components and even full-scale demonstration articles accomplishes several things:

1. It builds customer confidence in the new material.
2. It can reveal flaws in the manufacturing process.
3. It can reveal tooling changes that need to be made prior to production.
4. It can often detect design changes that need to be made before the production design is committed.

Materials development process

In the somewhat traditional materials development environment, the R & D engineer has generally worked very closely with the material supplier community to evaluate new materials, usually within a Materials R & D group, see Figure 1. Quite frankly, this makes the job of transitioning the technology to the internal customer, who is usually located somewhere else on the XXX Aircraft Project Team, somewhat difficult. Unfortunately, this arrangement often leads to confusion, miscommunication, mistrust, and sometimes even resentment. I call this a 'push'-environment, where the R & D engineer is trying to push their technology onto the programme. Quite frequently, the programme pushes back.

A superior materials development process is the Integrated Product Team (IPT) approach, in which the R & D engineer preferably sits with their customer on the programme and they develop the technology together, see Figure 2. If collocation is not feasible, then the R & D engineer needs to faithfully schedule periodic status reviews with his customer(s). Other key members of the IPT include the Materials Supplier and even the end-use customers, e.g. defence agencies, aircraft regulatory agencies and maybe even airline customers. This creates a 'pull'-environment, in which the customer pulls the technology onto their programme. In my opinion, an R & D engineer is truly successful when the customer embraces the technology and considers it theirs and not yours.

Integrated Product Team Frequent status reports

Figure 2: Modern materials development.

Start simple

In St. Louis, our initial application for a composite structure was a set of 50 flight-test rudders for the F-4 during the mid-1960s. The initial success with this programme gave us the confidence to build the F-15 empennage, i.e. vertical fin, horizontal stabiliser and rudders, out of composites. A few years later, the speed brake was converted to composites when the original metal speed brake proved to be too small. In the mid-1970s, the F/A-18 programme committed to even more composites and the highly-loaded, safety-of-flight inner and outer wing skins were fabricated out of them. In the early 1980s, when the original Harrier was redesigned, composites had matured to the point where both the wing skins and substructure were built from them, along with the forward fuselage and integrally co-cured horizontal stabiliser. Composite materials have proven very beneficial to military aircraft, especially US Navy aircraft, which must withstand the harsh environment of the aircraft carrier. The Chief Engineer of Naval Air Engineering once told me that he liked composites 'because they don't rot (corrode) and they don't get tired (fatigue)'. The real message here is that when you have a new and unproven material, start with simple applications, demonstrate early successes and then build on them with confidence.

Two examples - one structured and one not so structured

An example of a success story was the implementation of high-speed machining (HSM) on the F/A-18 E/F aircraft. Most technologists think of HSM as a high-productivity, low-cost alternate to conventional machining, which it is. However, the real driver for implementation on the F/A-18 E/F programme was weight reduction. By using HSM techniques, we were able to replace built-up sheet metal and machined assemblies with one-piece machined integral structures, see for example Figure 3. The current F/A-18 E/F contains over 100 high-speed-machined assemblies, saving significant costs and weight. There were several keys to the successful implementation of this technology:

1. A collocated multidisciplinary team of technologists was assembled to develop the technology, including experts in machining, cutting tools, vibration analysis and equipment.
2. A concerted effort was made to keep both the F/A-18 E/F internal customer and the US Navy customer up-to-date on the progress.
3. A development HSM cell was set up in the R & D laboratory where the process was developed.

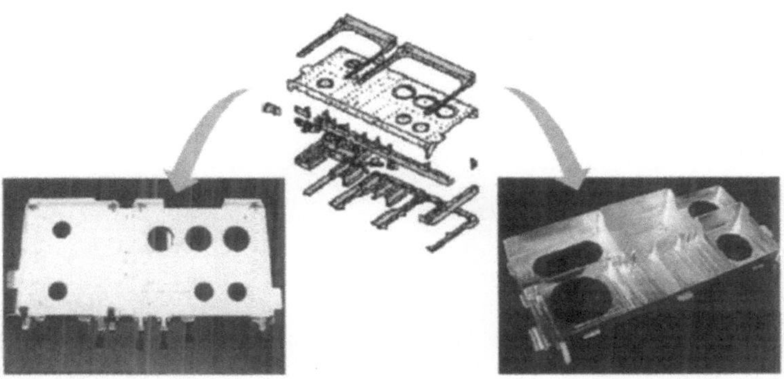

	Was	Now
number of pieces	44	6
number of tools	53	5
design and fabrication hr (tools)	965	30
fabrication hr	13.0	8.6
assembly man-hours	50	5.3
weight (lb)	9.58	8.56

Figure 3: Example of built-up assembly replaced with one-piece integral structure; F/A-18 E/F bay 4R shelf.

In addition, this development cell proved invaluable in building customer confidence in the process. Both the F/A-18 E/F internal and the US Navy customer could actually stand there and watch a simple HSM part be generated within a few minutes.

The second success story is a little messier. During the early portion of the flight-test programme of the AV-8B Harrier, the heat reflected from the ground back up onto the underside of the aircraft proved to be much hotter than anticipated. The heat was so intense that portions of the lift improvement devices (strakes) and the inboard portion of the lower flap skin, see Figure 4, which were fabricated from carbon/epoxy material, were showing visual signs of heat damage. Fortunately, we had been investigating fibreglass/bismaleimide for some secondary hot structure. Rather than convert the strakes and lower flap skin to titanium at a severe weight penalty, we embarked on a crash development programme to qualify carbon/bismaleimide to replace the carbon/epoxy in the hot areas. I tell this story not to illustrate that this is the correct way to implement a new technology. My point is that I have rarely seen new technology inserted on an aircraft programme that has passed certification testing (static and fatigue)... unless something is broken and the alternatives are limited. However, I will admit that in the past several years this paradigm is changing due to the much greater emphasis on affordability. If a good, sound financial case can be made to change to a new material, programme managers are much more amenable to making the change. Still, as a general rule, it is much easier to insert a new material or manufacturing process during the early portion of the programme before the design is frozen.

Figure 4: AV-8B Harrier conducting hover.

Conclusions

In closing, I will offer three simple recommendations to new materials development and integration:

1. Do your homework, i.e. know your materials and processes.
2. Live with your material supplier(s) and customer(s).
3. Start simple, demonstrate successes and build on confidence.

Response 2

Wings of silver, wings of gold: Money and technological change in the aircraft industry during the 1920s and 1930s

Marc L.J. Dierikx

Scriptura Research

To describe the rapid changeover from 'wood' to aluminium in aircraft construction, Schatzberg evokes the image of 'the owl of Minerva that only flies at dusk'. A myth is an appealing perspective to examine this process – indeed Schatzberg aptly chooses the imaginative to present his case. Is, after all, a popular definition of a myth not an imagined story that holds an element of truth? It is uncertain who first quipped 'money makes the world go round', but if one was to put another myth against Schatzberg's owl, this would be a likely candidate. We all know that the turning of the world is dictated by processes that have absolutely nothing whatsoever to do with the human invention of currency, yet if we perceive 'the world' as our own modern society of the 21st century, one would be hard pressed to come up with valid arguments against the crucial role that money has in our society.

But let us not divert to economic philosophy, and return to the topic that is under discussion here, i.e. changes in the use of structural materials in aircraft construction. Schatzberg makes an interesting and appealing case for the influence of 'culture' on the development of this specific materials technology. His explanations certainly hold elements of validity. On the other hand, Schatzberg's approach to what is called the social construction of technology has clear delimitations. One gets the impression that Schatzberg's 'fashion-conscious' engineers are working in a secluded environment, where only the faintest sounds of the larger society permeate. His aeronautical laboratories appear to be enclosed by high walls beyond which few things happen that are relevant to the content and outcome of engineering processes. In short, the story presented contains but a part of the full picture of what happened. Schatzberg creates a myth.

Historians are concerned with the analysis of past processes. Hence this contribution will not attempt to take up the very recent development of Glare here, or for that matter, its 'predecessor' Arall. Instead, we shall focus on aeronautical engineering in the 1920s and 1930s to demonstrate that aluminium constructions were not the outcome of a more or less 'logical' technological path along which organic materials were replaced by inorganic, man-made materials as the machine age progresses. Neither were they the result of an engineering culture in the way Schatzberg suggests. In doing so, the following attempts to demonstrate that historical analysis holds generalities that can be applied not just to past processes, but also for the present day and the future. To do so we shall examine the story of perhaps *the* proponent of the 'wooden' construction in aeronautics, Anthony Fokker.

To determine whether the change from wood to metal in aircraft construction was indeed linked to something like 'culture', or whether it was part of different, socio-economic processes, we need to go back to the origins of the use of metallic structures in aircraft in the first decade of the 20th century.

Experiments in metal

From 1897 the German scientist and entrepreneur Professor Hugo Junkers was involved in the research of various kinds of engines and diverse machinery at the Technical University at Aachen, Germany. Before that, Junkers had been involved in industry. After his study in mechanical engineering he founded a *Versuchsstation für Gasmotoren* (trans.: test facility for gas engines) in Dessau in 1890 together with a partner, Wilhelm Oechelhäuser. Five years later he began his own company, Junkers & Co., in which he produced gas-fired water heaters for domestic and commercial use. At Aachen, one of his colleagues, Professor Hans Reissner of the discipline of technical mechanics, experimented with aeroplanes – more or less as an (expensive) hobby. Reissner, however, soon ran out of money. After damaging his Voisin biplane, he applied to his colleague Junkers for funds. Junkers, after all, had become affluent as a result of the commercial success of his water heaters. In October 1907 Reissner proposed that he and Junkers should join forces in Reissner's aeronautical research. They agreed on a joint programme for technical research on flight. Reissner's Voisin served as a basis for experimental improvements to its design and construction. This took over a year. One of the areas on which the two scientists focused, were the machine's wings with their complicated structure of bracing wires. Despite these, Reissner found that the aircraft lacked sufficient stiffness in its construction, which was something he wanted to improve upon. He and Junkers studied the aircraft's deficiencies and came up with a partial solution to the lack of stiffness; their Voisin adaptation was to reconstruct the fuselage from welded steel tubing. Apparently, this solution was not the answer to the deficiencies of the machine, because the adapted craft crashed on a test flight in July 1909. Reissner then approached Junkers with a general design of his own for a monoplane. It was agreed that Junkers would design and construct the wings. The reason for this proposal was rooted in the past experiments of the two scientists seeking greater structural stiffness. To achieve this, Junkers proposed using a sheet metal wing covering, instead of the fabric covering that was customary at the time, see Figure 1.

Junkers' choice for metal had nothing to do with engineering culture. Indeed, certainly in aircraft construction, there was no such thing as a developed 'culture' at that point other than the widespread practice of hands-on tinkering, trying to improve upon a design during construction and on the basis of flight experience. What Junkers was looking for was a strong wing that would need less bracing wires – basically an engineering challenge. So, given the choice of materials, why did he come up with metal? To explain this, one needs to take into account Hugo Junkers' diverse interests in engineering – and in business. In his water heater business Junkers had accumulated a lot of experience with producing, cutting and handling thin sheet metal. Junkers had used sheet metal as a cover for his water heaters for years and was thus intimately familiar with the material. It was Junkers' belief that thin sheet metal might just have the characteristics that he was looking for without too much of a weight penalty over doped fabric. Besides, he had at his disposal the industrial facilities of his Dessau factory, equipped for cutting and handling sheet metal.

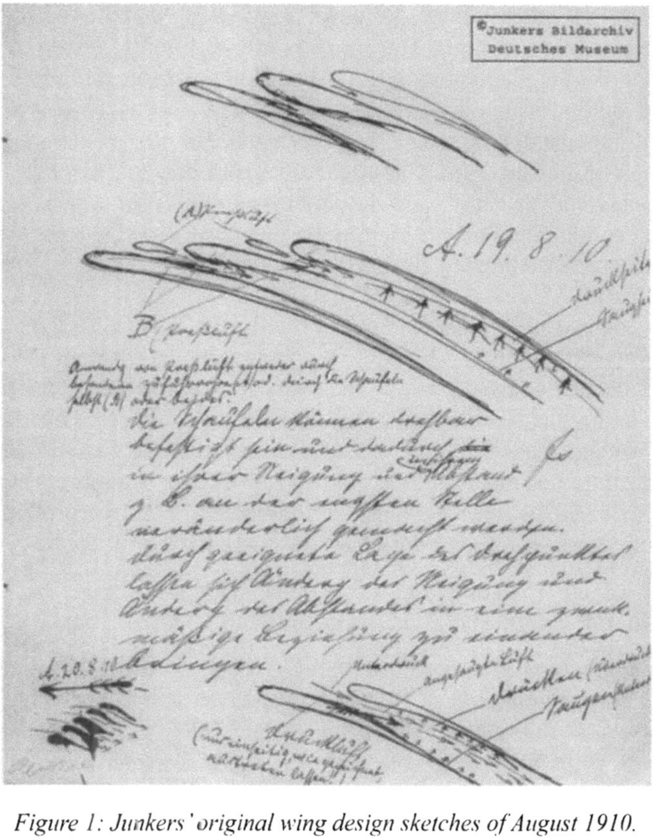

Figure 1: Junkers' original wing design sketches of August 1910.

Figure 2: The Reissner 'Ente' *with Junkers' corrugated wing in 1912.*

Things were not quite as simple as that, however. Laboratory testing proved that sheet metal, although stronger than fabric, would still fall short of expectations. Junkers found the sheet metal to have insufficient buckling strength for the envisaged use as wing covering. But Junkers, who divided his attention between the university and his own laboratory and construction company, did not give up easily. In what we would probably describe as a case of path dependency, he persisted, as a scientist, with his metal wing cover project. One of the things he found was that he could get the required strength if he used corrugated sheets. Thus the wings for the new Reissner aircraft had the usual fabric covering replaced with corrugated sheet iron – the same material Junkers used in his gas-fired water heaters and in various kitchen appliances his company developed.

But if corrugated sheet iron resolved the stiffness problem that Junkers and Reissner had set out to tackle, it again came at a weight penalty. In a further effort to resolve this, Junkers decided to use the newly discovered metal aluminium instead. It was in this experimenting that Junkers stumbled on what would become his trademark, i.e. aircraft with wings made from corrugated aluminium sheet material. In June 1910, after much testing and tinkering, he was able to deliver aluminium wing sheeting that had a thickness between 0.3 and 0.5 millimetres. The two professors now set about refining Reissner's design for what was going to be a monoplane, the Reissner '*Ente*' (trans.: Duck), see Figure 2. For this design Junkers' corrugated wing was essential, see Wagner [1] (62-68).

It was the combination of weight and stiffness requirements that started the use of, and research into, aluminium in aircraft construction – not unlike the start of the research into Arall and Glare over half a century later. Culture, or 'fashion', had nothing to do with it; as scientists and engineers Reissner and Junkers simply wanted to create a sturdy aircraft that could fly. A flying machine dictated that it should be light – aero-engines had little horsepower in those days and had an unfavourable power-to-weight ratio – and sturdy. Applying metal, out of which Junkers also built his engines and machinery, did just the trick.

Nonetheless, making aircraft from metal did not become fashionable for several decades to come. There were several reasons for this. First and foremost was probably the lack of formal engineering training that most early aircraft producers had. Contrary to the more theoretically and scientifically inclined ways in which Reissner and Junkers developed their flying machines in the university laboratory, the 'practitioners in the field' did not have many months available to spend on the development of aircraft structures exactly suited to their needs. The learning curve they faced was not theoretical, but empirical. They built aircraft based on acquired know-how, learning from each new idea that appeared to work. New aircraft types were developed in a time frame of weeks, months at the most. To give an example, the Dutch-born German aircraft producer Fokker produced 110 prototypes during the First World War – an average of one new model every two weeks. Practice, not theory dictated construction. The learning curve called for flexible work methods and materials that could be easily adapted – adapted by hand, if possible. Wood and fabric were such materials, and they remained standard in the rapidly changing world of aircraft development.

Secondly, money was of overriding importance to most aeroplane builders. Like any business, they needed to recover their costs, and make a profit if at all possible. This meant that costs had to be kept low. The market for aeroplanes, certainly before 1914, was very small indeed. Apart from piecemeal military orders, it consisted almost exclusively of affluent private individuals. This had an effect on materials choice. With aircraft engines fairly unreliable, aircraft were prone to be damaged after every few flights. Like in the case of the early automobile, it was imperative that their owners should be able to repair their own craft. This too dictated the use of materials that could be obtained widely, and that were easy to work with. Besides, with production runs being extremely small before 1914, it was not very practical to construct aircraft out of anything but materials that were widely available and which could be adapted easily with every new aeroplane built. The combination of these factors meant that wood and fabric became the standard materials for aircraft construction.

From metal to wood

The epitomical producer in combining the early construction techniques was Anthony Fokker.[1] Fokker, born in the Dutch East Indies in 1890, shared with many early aviation pioneers the fact that he was an autodidact to the profession. He had very little, if any, formal training in engineering. Around 1910, when Fokker first entered the world of aircraft construction as a rich man's son, building aeroplanes was not so much an engineering process or even a business activity, but rather an

[1] Much of the following is derived from Dierikx [2] (passim).

expensive hobby of a few people who shared a vision and the hands-on experience of how to make craft that could actually fly. Aircraft were developed on a trial-and-error basis. Ideas that flew – literally – were incorporated into the next design, weeks or months away. But even in those days, the outcome of ideas – aircraft – needed customers that enabled their creators to continue their uncertain route along the learning path of the new technology. Certainly for Fokker, this path seemed narrower than to others. After having flunked high school, his father had sent him from Holland to Germany at the age of twenty in order to become an automobile technician. Young Anthony Fokker, however, developed a different passion: aeroplanes. In the summer of 1910, less than a week after his arrival in Germany, he stumbled upon flying. His dream, supported by his wealthy Dutch family, soon became the construction of aircraft after his own designs.

Fokker's prime assets were his self-taught piloting skills. These he chanced, along with his life, on each new design that reached fruition, determining the fine-tuning of the machine on the basis of what the German's call *Fingerspitzengefühl.* He would later brag that he threw out each new engineering textbook as soon as he had taken a peek at it [3]. All the same, Fokker's method of working was that of most pioneers in a new area of technology, and differed little from that of his main competitors. But as a Dutchman among native German constructors in an extremely small market, *Fingerspitzengefühl* was not enough to make it in the world beyond the aerodrome. Always on the verge of bankruptcy, Fokker's small outfit remained a marginal affair in German aviation. By 1914 Fokker realised that if he wanted to survive as a constructor among the competition, he needed to safeguard his business interests. This meant he needed to sell more aircraft. And selling more aircraft meant securing orders from the German military. To effect this, he realised, he needed to make a jump up the learning curve and put out a machine that would be superior to those that his German competitors were constructing. If anything, such a jump would be expensive to make – probably involving the hiring of a qualified designer – and likely to require that he develop and try out new engineering ideas and practices that were far beyond his means, no matter how much money his affluent father and his family dared risk to invest in his business. Instead, he decided to rely on copying the results of engineering that had been developed elsewhere – a practice that was quite widespread in those days. In Paris he bought the wreckage of a high performance French Morane-Saulnier aircraft and took it apart. Fokker actually improved on the design, changing the construction of the fuselage structure from wood to welded steel tubes. With these adaptations Fokker's Morane-Saulnier copy had just the right combination of flexibility and rigidity needed for a performance aircraft. Its lightweight Gnôme rotary engine added to the aircraft's flying characteristics. Because Fokker did all this at a crucial point in time, i.e. only a few months before the outbreak of the First World War in August 1914, he happened to come up with the only highly manoeuvrable reconnaissance aircraft the Germans possessed just when demand for such a machine peaked. The road to success is often paved with chance. In the following months and years Fokker built upon the experience gained from his Morane-Saulnier copy, ascending to the top of the German wartime aircraft industry.

Junkers, by contrast, climbed the learning curve in a different fashion. For years he conducted tests in his large wind tunnel at the university. His goal was the development of a non-braced cantilever wing – a major step in aircraft engineering in those days. Junkers, building on the experience gathered from the Reissner '*Ente*', was convinced that a monoplane aircraft with a cantilever wing offered at once substantial weight savings *and* superior lift qualities compared to the 'conventional' biplane aircraft that were then in use. In Junkers' experience, such a wing should derive its strength from the use of metal construction. In May 1915 his researches had progressed far enough for him to invite a delegation of military officials to visit the Junkers' facility in Dessau and witness the progress that had been achieved. To his disappointment, Junkers found the military sceptical of his designs. An aircraft made from metal, created in a company that 'made kitchen utensils' was regarded with suspicion. In spite of this, Junkers managed to obtain a contract for a test machine, the Junkers J1. Because Germany did not possess any supplies of aluminium, it was decided the aircraft should be built from steel (tubing) and thin sheet iron, materials that Junkers was intimately familiar with. On December 12, 1915 the aircraft was ready for its first flight test. Subsequent testing showed the machine to be faster *and* have better climbing performance than the conventional biplane fighter aircraft (Albatros D.II) of wooden or mixed construction that were in use with the German armed forces at that point, see Wagner [1] (79-86). However, flight-testing and fine-tuning the aircraft to Junkers' high demands as a scientist and an engineer took over a year. By that time, the German military, pressed by battlefield needs, had become anxious: 'Would Junkers be able to mass-produce such a highly engineered aircraft?' And, moreover: 'Would such an aeroplane really be better than the conventional wooden and fabric machines operated over the front?' In December 1916 the German military forced a joint venture between Junkers and Fokker to ensure optimal operational characteristics, and bring on board experience in series production: the *Junkers-Fokker Werke AG Metallflugzeugbau.*

Although the collaboration between Junkers and Fokker did not lead to mass production of Junkers' aircraft, its consequences in the long term were considerable – certainly for Fokker. Fokker used his association with Junkers to obtain detailed knowledge about Junkers' design for a cantilever wing, and then used that information to produce a cantilever wing of his own. Fokker, sticking to his customary practice of working with wood, built *his* cantilever wing from wood, instead of metal. Even so, the Fokker wing seemed to represent an infringement on Junkers' cantilever wing patent. Junkers was furious, wishing to end his association with Fokker at the earliest possible moment. Legal proceedings over the infringement upon Junkers' patent lasted until 1940. The case remained inconclusive, however. For Fokker, as in the case of the Morane-Saulnier design, mere copying was not his style. Instead, he used his association with a Swedish engineer also working in Germany, Villehad Forssmann, to produce a plywood veneer wing cover for his own version of the cantilever wing. In this wing, Fokker's plywood replaced Junkers' corrugated metal. In some ways, this actually represented a step forward over the Junkers wing cover as far as strength was concerned, because plywood, with the wood grain of the various layers running at right angles to each other, offered even better buckling strength than Junkers' corrugated sheet metal. Fokker used the new wing construction in his designs for the cantilever-winged fighter aircraft that dominated Germany's air war effort in the closing stages of the war: the biplane D.VII, which featured semi-cantilever wings with a plywood leading edge, but further covered with fabric, and the monoplane D.VIII, which featured a cantilever wing with full plywood covering. Both fighters entered service with the German Air arm in the spring of 1918.

When the war ended, Fokker elaborated on the experience he and his design team had gathered in the construction of cantilever wings with plywood covering and used these as a basis for the design of his series of transport aircraft. From the design of a long-range reconnaissance aircraft, the F.I, Fokker's team developed a high-wing, single-engine transport aircraft, the F.II, which subsequently served as a model for subsequent developments. Like Fokker's wartime aircraft, the F.II had a fuselage made of welded steel tubing, basically covered with fabric. But the war brought Fokker more than just experience with aircraft development. The end of it, and the subsequent revolution that enveloped Germany in the months after the Armistice in November 1918, traumatised him. Much of the fortune that Fokker had amassed during the war, and for which he had worked incessantly ever since his days of hardship came to pass in August 1914, evaporated in the immediate postwar months. Of the thirty million Marks he estimated he had accumulated, he was only able to recover about 25 percent. With that money, Anthony Fokker crossed the border into Holland in February 1919 to set up his company anew.

In the subsequent years, the F.II would become the conceptual model for the development of new, ever-larger transport aircraft at the Fokker factories. Although Fokker left much of the actual technical work to his small team of designers and engineers, he did decree that subsequent developments should derive from the standardised concept that the F.II represented. The reasons for this were surprisingly simple. For one thing, increases in size based on the basic airliner concept could be effected relatively quickly, and at comparatively little cost. They depended primarily on the availability of more powerful engines and the application of aerodynamic improvements and allowed Fokker to continue to incorporate insights gained from practical flight experience. Fokker's most expensive civil aircraft to appear for over a decade, the F.VII airliner of 1924, cost no more than $34,750 to develop, see Dierikx [2] (211). Other aircraft cost (considerably) less. Not one to lightly forget what had happened to his capital in the aftermath of the First World War, Fokker took extreme care to keep costs down. After all, Anthony Fokker was the sole shareholder and investor for his Dutch postwar company. Fokker's Dutch company made most of its profits from the export of military aircraft to such customers as the Soviet Union and the clandestine German rearmament programme. Other customers for his military aircraft included European governments that faced the need to modernise their air force, but were at the same time cost-conscious and generally inclined to favour indigenous producers unless Fokker managed to substantially undercut them. By the early 1930s military expenditure dried up even further, as the economic crisis set in and international negotiations towards disarmament appeared to be making progress.

The market for world passenger transport aircraft was precariously small in the 1920s, and besides all of Fokker's potential customers for such aircraft were operating at considerable losses, curtailing budgets that might otherwise have been available for the purchase of new aircraft. By reducing the cost for research and development of new aircraft to a minimum, Fokker was able to get by on comparatively small series production of each new aircraft type. Doing so, he was also able to customise his aircraft types to the specific wishes of his clients. At the same time, working with conventional materials allowed for easy repairs in operational use. Fokker was extremely successful at this. Indeed, the Fokker company became the world's leading producer of civil airliners. In the early 1920s his business interests spanned Germany, Holland and the Eastern seaboard of the United States, with Fokker aircraft factories in each of these countries. For Fokker's American operations, several new production units were envisaged, giving his company new footholds in the Midwest (Kansas) and in California. For Anthony Fokker personally, his parsimonious way of running his various companies – those in the United States, set up at the insistence of Air Force General Billy Mitchell, were funded with American private capital – paid off. Exports and licence production contracts for Fokker aircraft were such that by the mid-1920s Fokker's trimotor development of the original F.VII model set something of a design standard throughout the industry. The F.VII-3m of 1925, see Figure 3, retained the customary Fokker design characteristics and had been adapted from the F.VIIa model at little more than the cost of installing two additional Wright Whirlwind engines. In the summer of 1925 the first aircraft was built within eight weeks of Anthony Fokker's initial telegraphic instructions, see Dierikx [2] (95-96).

Figure 3: Anthony Fokker posing with the original Fokker F.VII-3m in 1925.

With the success of the trimotor aircraft there was little need for strategic reorientation. Besides, as owner-director Fokker typically operated on a short-term perspective. Yet the second half of the 1920s brought fundamental changes to the industry, certainly in the United States – changes that Fokker did not recognise in time. New legislation aimed at stimulating the development of a continental air transport system – the 1925 Air Mail Act and the 1926 Air Commerce Act – heightened public awareness of the potential of aviation. A year later the Lindbergh-flight appeared to prove that high expectations were justified. In a booming economy an upsurge of investments in aviation shares occurred on the Wall Street stock market. Capital influx into the aircraft industry was rising rapidly, with a number of major industrial corporations pouring in money and resources. This changed the industry forever.

From slide rule to dollar rule

In the late 1920s venture capital and market expectations brought a fundamental shift to the way in which aircraft were being designed and built in the United States.[2] Hitherto, the industry had maintained many of the characteristics of small-scale entrepreneurialism, with empirical approaches having the overtone in aircraft development. The leading producers were all headed by men who had come to the industry through their personal experience as pilots and constructors. Now, for the first time, increased funding and expectations of a growth of the market for air travel made a more fundamental, 'scientific' approach to aircraft design possible. The emphasis in technological innovations in the aeronautical industry shifted. Although, in the 1920s, much of the scientific debate in aerodynamics centred on drag reduction in airframe design, the major practical innovations in aviation had come from the development of more reliable, lightweight air-cooled engines. In the US these culminated in the Wright Whirlwind engine, introduced in 1925. With the air-cooled engine in place, development efforts received a new impulse. The problem with the air-cooled engine was its relatively large frontal area, which increased drag resistance. In the second half of the 1920s research at institutions like the NACA Langley Laboratory shifted to the development of an engine cowling for the radial air-cooled engines that would solve the drag issue [7].

But why would designers leave wood? The process did not come automatically. Schatzberg rightly reconstructs that metal aircraft were neither cheaper, nor better than their contemporary 'wooden' counterparts [8].[3] As the case of Junkers explains, metal was just a *different* structural material. The first steps along the path of

---

[2] By comparison, aeronautical development in Europe stagnated because of a general lack of funding. In his standard work on the history of the aeronautical industry in France, Emmanuel Chadeau aptly summarised the situation as '*des capitalistes au petit pied*' [4]. In Weimar, Germany, where finances were geared towards research rather than industry, funding declined as the effects of the economic crisis set in [5]. For a general evaluation of the situation in Britain, where until 1930 much of the effort was concentrated on the development of airships, see Fearon [6].

[3] See also the keynote lecture by Schatzberg in this volume.

designing aircraft more suited to series production were made with wood. They came about on the threshold of a new era in air transport. In 1926 the Loughead brothers combined with designer Jack Northrop in their Hollywood garage to build a new type of airliner, incorporating the various innovations in aero-engine construction and aerodynamics that the first half of the 1920s had brought. The small scale of their outfit dictated the material used for their product; the Lockheed Vega of 1927 was built from wood. It incorporated the characteristics that were to make up the typical modern transport aircraft of the 1930s and combined the air-cooled radial engine with the advances made in aerodynamic design. It was the first production aircraft with the NACA cowling.

In comparison to the use of air-cooled engines and more advanced aerodynamics, metal came in late. The reason was that metal, a less flexible material than wood, required expensive tooling to be able to use it as a construction material. The use of metal held economic advantages when used in industrial series production – but only then. The advantages of metal were also its weak points; without the prospect of substantial series production, the use of metal was financially impossible. Aircraft producers that adopted metal, like Ford in the United States, did so for reasons of competition. To succeed in a market already carved up, they needed to offer something different: *metal* aircraft. But since metal was just *a solution* – not *the solution* – to constructional issues of strength and offered no real advantages over 'wood', something like a PR-campaign was necessary to present the metal aeroplane as something *better*, or more modern than the existing aircraft. If there was such a thing as a progress ideology, as Schatzberg maintains, it was likely orchestrated by the small band of producers who had enough money and industry – like Junkers, Ford had, of course, ample experience in working with sheet metal – behind them to even consider metal constructions. From an economic and production perspective, metal required market growth. This appeared after 1929, when the American air transport market developed as a result of the Air Commerce Act, which opened up numerous new routes under the patronage of the US Post Office. These new routes required small- to medium-size aircraft to operate. As profits were non-existent to marginal, operating the most effective equipment was vital to the operators. It so happened that market expansion and capital influx into the industry coincided with the spread of new approaches to aircraft design. This made metal suddenly an attractive alternative construction material.

Fokker, the leading constructor, was moving in another direction, however. He interpreted the Post Office's policy in a different way. Fokker expected *transcontinental* air transport to blossom before new regional 'thin' routes would be opened. For the transcontinental route, on which he expected substantial traffic increases, he had his designers develop the giant F-32 airliner: a very large, luxuriously fitted four-engine machine capable of carrying up to 32 passengers over long distances, see Figure 4. The F-32 had more than double the capacity of the average passenger airliner of the period. Adhering to his customary construction practice, Fokker expanded on the earlier designs for three-engine aircraft, adding size plus an additional engine for the necessary extra power needed. Why such a big machine? Fokker rightly surmised that to carry the transcontinental traffic efficiently, he needed to come up with an aircraft that would offer substantially lower costs per seat-mile. Lower costs per seat-mile meant having a larger aircraft, an economic wisdom that has remained in the airline industry, viz. the Boeing 707, Boeing 747, and Airbus A380. From 1928 the American Fokker corporation, then at the summit of its market leadership, devoted all of its resources to this single project for the F-32. Fokker was not the only one in drawing these conclusions as to the importance of costs per seat-mile. The Curtiss Condor two-engine double-decker, capable of carrying some eighteen passengers, was essentially built along the same line of conclusions, primarily catering for an expanding market on the transcontinental routes.

Figure 4: Festivities on the occasion of the acceptance of the first Fokker F-32 by Western Air Express in March 1930.

American air transport developed differently, however. Opening new routes turned out more important for traffic growth than operating coast-to-coast flights. To serve the expanding network, smaller aircraft were needed than the giant Fokker was working on. Not just smaller, but also – and primarily so – more efficient aircraft. If upstart airlines of the late 1920s, fuelled by the aviation boom on the stock market, were to have a chance of survival, they needed equipment that offered substantially lower operating costs so that they would be able to fly and develop the new thin routes. Such new, more efficient aircraft needed to differ from the dominant technology of the day as epitomised by Fokker. If Fokker went for increased size – with basically the same technology that had brought his company to the fore of air transport –, his competitors needed to go for technology. Their aircraft would not be custom-developed, but emerged from drawing boards and wind tunnel models. The new generation of aircraft combined the wind tunnel experiments in more aerodynamically efficient designs. It was to help this process along that the NACA developed the aforementioned cowling for air-cooled engines.

Of course Fokker knew about these experiments, yet early tests with a military Fokker C-2 Trimotor equipped with NACA cowlings showed disappointing results – later shown to originate from Fokker's practice of placing the engine nacelles underneath the wing. He surmised that the new research was moving along a dead-end street, and saw this as a confirmation that the immediate road ahead lay in size and comfort rather than in technology and efficiency. When Fokker was proved to be wrong, in the failure of his F-32 airliner, the company had no other viable projects on the drawing board. Looking for a scapegoat the new owners of the Fokker Aircraft Corporation, General Motors, decided that Anthony Fokker had to go, so that the company could make a fresh start.

But why did the competition manage to effect the changes they incorporated into their new designs? First and foremost, the money that was pouring into the aircraft industry was put to use by hiring qualified designers and engineers from places like MIT and Caltech. Upstarts like Lockheed's were lucky in the sense that their first products, incorporating some of the new technologies, hit the market at just the right time to be successful. This brought in extra money, while at the same time giving the incentive to head further along this new road. If the market continued to expand in the same fashion, larger numbers of 'standardised' aircraft might well be sold. Building larger numbers made research and development investments in metal worthwhile.

But investments were exactly what Fokker shunned. Aircraft designs, whether in wood or in metal, which departed from the customary practice of empirical improvements and upscaling of existing models would require serious investment. On the one hand, Fokker did not wish to risk money, on the other hand the American business partners he had brought on board were in the business for profits, not as providers of venture capital per se. In his preference for dealing with people, not institutions, Fokker had associated himself with financiers whom he knew personally and who were, like Fokker, ready to invest but extremely anxious about losing their money. Besides, Fokker did not have it easy in the US. Without the trusted team of designers and constructors that was available in Amsterdam, and without the scientific assistance of the Dutch government bureau for the study of aeronautics, the RSL, which would normally check and advise on all drawings and calculations made for new aircraft, Fokker was forced to be conservative in his US plant. Without the RSL safety net, and with only a small staff in the design and construction bureaux, Fokker aircraft that emerged from the American factory at Hasbrouck Heights, New Jersey, were even more conservative than contemporary models developed in Holland. At the same time, aircraft like the Fokker Super Universal of 1927/'28 and the F-10 of 1929 showed Fokker at his old-time best: tinkering on the basis of earlier designs finished in Holland. Even so, Fokker did manage to capture a large share of the emerging American market for commercial airliners with minimal investment in fixed assets and minimal investment in R & D.

He must have laughed at efforts undertaken in the Stout and Ford workshops to come up with costly designs of airliners constructed from aluminium, which offered few, if any, operational savings over Fokker's competing aircraft built from wood and steel tubes. Fokker, before all else, was in the business for the money. If he realised he had become trapped in path dependency then he did not care about it. From the typical short-term perspective on which Fokker operated, all looked fine.

But was it? Fokker depended on hiring and training precision wood workers, welders and craftsmen. Finding people with the precision skills needed for aircraft production was not easy. When Fokker opened up his new facility in Wheeling, West Virginia, in 1929 – being attracted there by local business groups – difficulties were encountered in hiring qualified craftsmen. Just how bad this problem was, was evidenced by the famous Knute Rockne crash on March 31, 1931. For starters, there were distinguishing factors that set the F-10A, the type involved in the crash, apart from Fokker's earlier aircraft types. It was the first Fokker model that was entirely designed and built in the United States. Up to then, Fokker had part of the design process and the wing production done in Holland, and imported all wings from his Amsterdam factory. But there was more to the wing of the F-10A. Designed in the empirical fashion in which Fokker built all his aircraft, the F-10A featured a substantially larger wing than earlier Fokker types. The wing, whose basic design was copied from that of the Dutch F.VIIb Trimotor, had a span that was 2.4 metres longer than that of the F.VIIb. This was, however, one extension too many, for the larger wing of the F-10A was prone to flutter, especially in extreme weather conditions. This made the F-10A a difficult aeroplane to fly, quite contrary to Fokker's earlier types. But not only was the F-10A, delivered from December 1928 onwards, the first Fokker design to have its wings built in the USA, it was also the first aircraft to be produced at the new Fokker plant in Wheeling. The combination of these circumstances contributed to the disaster; the Fokker F-10A Trimotor involved in the infamous crash was proved to suffer from imprecision in the wing construction, resulting in bad gluing of the various internal parts in the wing, which, in combination with flutter and bad weather, proved fatal [9]. All F-10A aircraft were grounded after the Rockne crash.

Without the crash, Fokker's heyday would have been over anyway. This was not because of the construction material per se. Subsequent designs for and successes of wood-built aircraft like the de Havilland Albatross, the Mosquito, the Hornet, the Westland Whirlwind, and even large transports such as the Messerschmitt Me 323 and – to an extent – Howard Hughes' 'Spruce Goose', proved that wood remained a feasible construction material for at least another two decades. During the Second World War, Germany's Arado Ar 234 '*Blitz*', the world's first true jet bomber, featured wooden wings. But by the end of the 1920s, Fokker had manoeuvred his company onto a dead-end track. Shunning investments in research and development, the empirical approaches to aircraft development and construction that Fokker used could go no further. On the other hand, Fokker's policy made business sense. The risks of developing aircraft that incorporated the latest technology were very high indeed. To recover the investments, large series of aircraft had to be sold. Douglas, using \$307,000 from a revolving bank credit on the DC-1/DC-2 development, needed to sell at least seventy-five just to break even at the relatively low price of \$65,000, as Douglas had to remain within the price range that the airline industry could afford, see Ingells [10].[4] For comparison, Fokker's F-10A sold at \$67,000 [11]. Fokker was simply not willing to risk that kind of money, and other old-time entrepreneurs like him often lacked the means. The financial development of the aircraft business from 1927 onwards made that the competition suddenly found itself ideally placed to sell increasing numbers of its (metal) aircraft and take over market hegemony.

After the American débâcle with the F-32, Fokker further retracted towards low-cost development of new types. This showed in the R & D figures at his Amsterdam plant, see Figure 5. If the single-engine F.VII of 1924 cost \$35,000 to develop, the F.IX Trimotor of 1929 cost half that amount, i.e. \$18,000. In 1932 the F.XII Trimotor was again cheaper to develop at only \$6,600. Such figures indicated a company in trouble, retracting towards very small series of aircraft custom-developed 'on a shoestring'. *The* extreme example of this process was the F.XVIII Trimotor that Fokker developed for KLM. It was derived from the F.XII at a mere \$481.25 in research and development at a time in which Douglas was getting ready to spend 770 times that amount on the DC-1/DC-2 design.

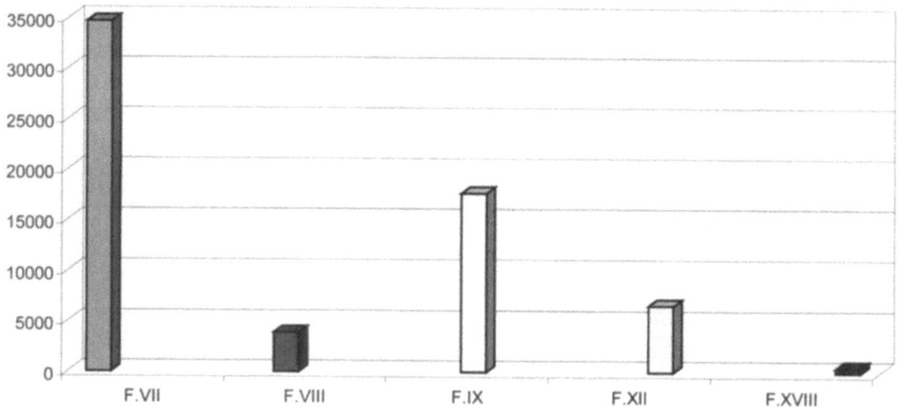

Figure 5: Civil aircraft development costs at Fokker in 1924-1933 (US dollars).

Conclusion

It is questionable whether there was a true changeover from 'wood' to metal around 1930. In practice, the two construction technologies had coexisted for decades. During those decades of coexistence 'metal' was not a challenge to 'wooden' construction – and for good reason. 'Wood' provided a relatively low-cost, flexible material, eminently suited to the type of practical engineering that was dominant in

[4] Ingells refers to sources at the Douglas Company.

the industry of the day. The aircraft business was a small-scale industry. Until the second half of the 1920s the industry was also dominated by the entrepreneurial pilot-constructors who had personally put up the money for their companies. To build aircraft in metal, most producers lacked the knowledge, experience and – above all – the means. Materials other than wood simply made no economic sense, except for a few well-funded military projects. Only when, in the second half of the 1920s, serious corporate investment in the aircraft industry became fashionable in the United States, did the industry enter into a new phase in which the various theoretical and practical notions that had been accumulated, could be combined in designs for a new generation of aircraft. In the United States the aviation industry witnessed the rise to prominence of various hitherto minor producers, who happened to be well suited to combine money and knowledge into designs that were specifically aimed at outperforming the hitherto dominant 'Fokker' types of airliners. Indeed, the way to compete with the dominant technological standard of the day was to offer something radically new. Airline operators were persuaded to buy these new 'standardised' aircraft because they offered far better operating economics than the existing types. In the case of the DC-2, operating costs were some twenty percent lower than those of contemporary Fokker models [12]. Added to that was the circumstance that the dominant producers of civil transport aircraft in the United States, Fokker and Curtiss, both aimed for a market development that did not come about – and therefore produced aircraft no-one would buy. In summary, economics, not engineering fashion, was the dominant factor in the changeover from one technological regime to the other.

References

[1] Wolfgang Wagner, *Hugo Junkers: Pionier der Luftfahrt – seine Flugzeuge*, Bonn: Bernard & Graefe Verlag, 1996.

[2] Marc Dierikx, *Fokker: A Transatlantic Biography*, Washington DC: Smithsonian Press, 1997.

[3] Doree Smedley and Hollister Noble, 'Profiles: Flying Dutchman', in: *The New Yorker*, February 7, 1931: 20-24.

[4] Emmanuel Chadeau, *L'industrie aéronautique en France, 1900-1950. De Blériot à Dassault*, Paris: Fayard, 1987: 204-ff.

[5] Helmuth Trischler, Luft- und Raumfahrtforschung in Deutschland 1900-1970. Politische Geschichte einer Wissenschaft, Frankfurt am Main: Campus, 1992: 142-173.

[6] Peter Fearon, 'The growth of aviation in Britain', in: *Journal of Contemporary History*, 20, 1985, no. 1: 21-40.

[7] James R. Hansen, *Engineer in charge. A history of the Langley Aeronautical Laboratory, 1917-1958*, Washington DC: NASA, 1987: 123-132.

[8] Eric M. Schatzberg, *Wings of wood, wings of metal: Culture and technical choice in American airplane materials, 1914-1945*, Princeton: Princeton University Press, 1999: 111-112.

[9] Aeronautics Branch US Department of Commerce, Accident Report, April 1931, in: National Air & Space Museum, Washington DC, Rockne Crash File.

[10] Douglas J. Ingells, *The McDonnell-Douglas Story*, Fallbrook CA: Aero Publishers, 1979: 46.

[11] Fokker Aircraft Corporation of America, sales brochure, April 1930, in: Aviodome Archive, Schiphol, the Netherlands, Fokker file.

[12] Minutes KLM Board Meetings, June 26 and July 28, 1934, in: KLM Board Papers, Amstelveen, the Netherlands.

Response 3

Fibre metal laminates: An evolution based on technological pedigree

Leo J.J. Kok

Bombardier Aerospace

The development of fibre metal laminates (FML) as a class of structural materials is the culmination of the coalescence of various well-developed technologies in the aircraft industry, with the basis of some of them dating back to the early days of flight. The process of bonding thin metal laminates reinforced with pre-impregnated fibres spliced together to make large fuselage panels has arisen from previous lessons and the economic drivers of airframe production, airline operations and shareholders expectations. The anthem of 'higher faster farther' applied during the development of the Comet 4 still applies today – with one caveat of course, i.e. reduced cost!

'The greatest difficulties lie where we are not looking for them.' – Johann Wolfgang von Goethe

The early days (recent history for some)

The use of construction techniques in the aircraft industry has largely been based on the innovative use of resources and practices that were readily available. The 1903 Wright Flyer embodied canvas, spruce, metal, fasteners, glue and paper and proved to be more than somewhat successful, albeit from a technical point of view. This first aircraft relied on technology from other industries and their practices, especially the bicycle. As designs developed, the use of wood adhesives became more prevalent, again a practice borrowed from another industry, in this case furniture manufacturing.

Tinkering with wood and metal

The world's first all-metal aeroplane, the Junkers J1, flew in 1915 using a stressed-skin construction in the form of corrugated aluminium. In 1919, Alcoa started producing sheet for aeroplane fuselages and wings [1].

When the United States of America entered World War II in December 1941, it was faced with an uncertain supply of strategic materials needed to produce large quantities of military aircraft. A solution forwarded by the Edward G. Budd Manufacturing Company of Philadelphia, Pennsylvania, the manufacturer of munitions and railroad rolling stock, was the Budd Conestoga, which made extensive use of stainless steel. After its first flight on October 31, 1943, this aircraft was delivered to the USN in March 1944. It later crashed during testing and the test pilot swore that the plane's stainless steel construction saved his life. The flying characteristics of the RB-1 were poor and problems with the use of stainless steel further delayed production and caused the price to rise. In any event, an adequate supply of aluminium and the availability of the C-47/R4D resulted in the USAAF cancelling their order for this aircraft and the USN reducing their order from 200 to a total of 26. All in all, the Budd Conestoga is not a well-known affair in the annals of aviation history [2].

More widely known is the very successful de Havilland solution to the same problem using wood. The result was 6,710 Mosquito bombers, fighter-bombers and night fighters used extensively during WW II. Not only did this allow for increased production of other planes, a significant amount of hot-bonding research was done in the aircraft industry, a technique which had lain dormant for some 30 years. Work at Hatfield, UK, in the mid-1930s focussed on the use of laminated wood veneers of birch, hickory beech or acacia arranged longitudinally to have high specific strength and stiffness for use in wing spars. Such a technique of fabrication is shown in Figure 1 with a stressed-skin wing shown in Figure 2 [3]. This research led to widespread use of synthetic glues for wood on the Mosquito production line with the classic picture of lifting the half shell moulds shown in Figure 3.

Figure 1: Wood laminating at de Havilland.

Figure 2: Stressed-skin work at Hatfield.

Figure 3: Lifting the Mosquito bomber carapace off the mould.

The advent of metal-to-metal bonding

An early use of Redux bonding in aircraft construction was on the Hornet single-seat fighter as recounted by Moss in 1949 [4]. A design dilemma arose in that the large wooden boom required for the wing spars took up much valuable fuel volume. The solution was to use an aluminium alloy extrusion for the spar cap, but still use wood for spar webs and compressive booms. Later, Redux bonding was also used on floorboards of Vikings and Viscounts wherein top-hat stringers were bonded under plywood floors. It was then natural that the decision to use Redux bonding on the Comet would follow in 1945 [5]. Meanwhile, at the Bristol Aeroplane Company and other places, autoclave-bonding technology was being developed with excellent results. It would take some time for autoclaves to grow larger. Around this time Fokker started to introduce Redux bonding into its operations, with investigations carried out by the NLR with respect to bonded structures for static and fatigue testing. As a result, a complete bonded wing of a Fokker-designed S-12 primary trainer was designed and manufactured. We can see a glimpse of the shape of things to come if we look at the Comet production methods. Figure 4 shows de Havilland's then newly acquired heat press capable of forming a 42' x 9' wing component, an example of which is shown in Figure 5. Large flat components are much simpler to produce than curved fuselage shells, as we will see. Concerned with the size of autoclaves, the design and production departments devised a simpler method to allow for Redux bonding in situ on fixed tooling. Examples are shown in Figures 6, 7 and 8, with a finished section showing the noticeable absence of rivets in Figure 9 [6]. Of historical note and significance to this conference is the debut of glass fibre on the Comet for structural, non-flight-critical components [6].

Figure 4: Wing plank consolidation press.

Figure 5: Comet Redux-bonded skin/stringer panel.

Figure 6: Pressure bar on mould tool.

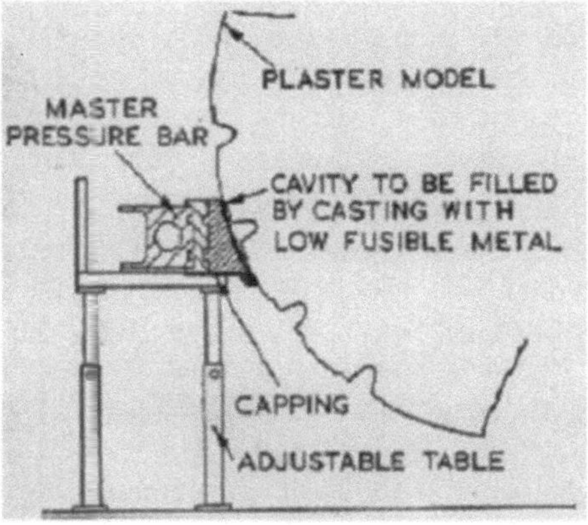

Figure 7: In-situ Redux bonding jig.

Figure 8: Forward fuselage mould tool.

Figure 9: Completed forward fuselage section.

Even then, it was known that the loads on joints were not uniformly distributed, but tended to concentrate near the ends of the joint. To alleviate this on thicker structures, tapered strap ends were used in the joint design [4]. With this design consideration it was still possible to achieve higher loads on bonded joints while eliminating the stress concentrations of rivets. The analytical details of this would be well documented under the Primary Adhesively Bonded Structure Technology (PABST) programme in the mid-1970s and reported by Hart-Smith et al. [7] some 20 years later. What was evident was that the structures designed for bonding achieved high strengths. Van Beek reports a 35% increase in strength under compression of a Z-stringer panel compared to riveted one [8]. Parker attests to up to 40% increases in compressive strength of bonded structures [9]. This compares well with compression data generated some 30 years later for a DASH-8 I/B flap of Arall 2 of an angle-section spar cap of 25% higher load [10]. Stiffened shear panel strength increases of 7-20% are reported and can be essentially attributed to the alleviation of the stress concentrations due to rivet holes [11, 12].

In remarks to a paper delivered to the Royal Aeronautical Society, by H.J. Pollard in 1953, H.B Howard from the Ministry of Supply mused that a 'combination of sheet metal and plastic laminate might be more efficient than either separately' [13]. The notion of combining materials together to form structural elements had already been seeded into the industry, albeit at the time the focus was directed at asbestos fibre mats, Durestos, which was relatively cheap and in vast supply.

After the War, as the aerospace industry sought to rebuild, practices in Europe and North America diverged somewhat. The readily available light-metal working machinery in Europe, lead to consolidation by way of bonding as the choice for aircraft construction. In North America this was also looked at, but the availability of then large amounts of capital that was placed into large extrusion and plate rolling mills progressed to machining of wing skins and chemical milling of fuselage panels followed by riveting [14]. As an example, for the Avro Arrow, a 3300 lb aluminium billet goes on and a 290 lb wing skin comes out at was then state of the art [15]. This became the prevalent practice in North America, debuting on the Boeing 707, see Figure 10 [16]. As part of the Commonwealth, and with the aerospace community being a little smaller, Canadian industry had access to both British and US technologies. Avro Canada broke new ground in bonding technology with its application of magnesium alloys bonded to aluminium structures to very tight dimensional tolerances and under the severe thermal environment of supersonic flight.

Figure 10: Boeing 707 fuselage skin riveting.

In spite of the lack of a widespread use of bonding in US aircraft programmes, research continued on bonding technology, notably at the Forest Products Laboratory of the United States Forest Service. Some thirty years earlier in 1920 the lab had done its first work for the aviation community in a contract for the NACA on the use of glues in aeroplane construction [17]. The work, circa 1950, developed the now famous FPL etch (sulphuric acid / sodium dichromate solution) and anodise with phosphoric acid used extensively in the industry [18, 19]. A lesser-known fact is that the team of Black and Blomquist stumbled upon a concept fundamental to the development of fibre metal laminates 30 years later. In their work, results were reported on the development of metal-metal bond adhesives up to 600 °F (315 °C) with great success using a dry tape supported on a glass mat base. They realised that joint strength is closely related to the physical properties of the metal (adherents) and the cured adhesive and to resultant stress concentrations in the joints under test. The researchers noted that:

> *'Joint strength might be improved appreciably by the use of supporting fabric in the adhesive bond. In subsequent tests made on metal lap joints bonded with a tape adhesive of a woven glass-fiber cloth impregnated with liquid FPL-710, the immediate strength of joints at 600 °F was materially improved.'* [20]

Subsequent reports suggested that the most promising binder material was found to be a 0.010" thick Owens Corning fibreglass mat, S11Mo1. In looking at ageing, producibility and peel resistance the team used the 1950s-version of FMLs as a test joint, see Figure 11 [21]. Admittedly, at the time the adherent was 0.032" thick. Laminating thinner sheets and their beneficial fatigue properties came to light a decade later with work done at the Alcoa Research Labs at New Kensington PA, and reported by Kaufman [22, 23]. Aside from the technical significance, this site would regain attention in the mid-1980s as the production centre of FMLs. Work under Johnson [24, 25, 26] at NASA Langley sought to quantify the effects of laminating thin metal alloys. Further work in Europe, especially at the Delft University of Technology by Schijve and Lipzig, reported similar findings on still thinner laminates [27, 28, 29, 30].

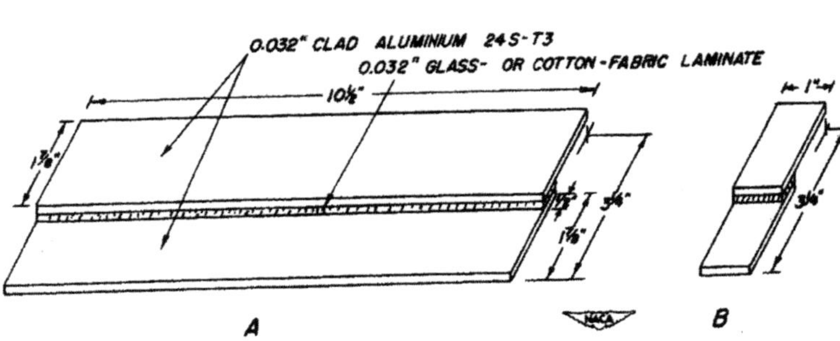

Figure 11: Test specimen for adhesive/plastics evaluation to clad aluminium.

Fibre metal laminates revisit history

In 1981 the field of fibre metal laminates officially took off with the granting of a patent for the concept of Arall [31]. It has been pointed out that work in the US had looked at using titanium whiskers as reinforcement, but the author could find no record of that work. In the litigious American society the Arall patent will therefore stand. Adhesion and environmental effects were examined in detail at the Delft University of Technology [32]. Of note is a reference to the work of Lekhnitskii some 30 plus years earlier on birchwood laminates and the analysis methods thereof [33]. Methods of Paris and Sih have also found applicability to this class of materials [34].

Coming out of the lab and into the aircraft industry work at Fokker [35] and de Havilland [36] spurred further interest. This was natural insofar as both companies were the only two remaining still using extensive metal-to-metal bonding in their products for primary structure.[1]

Prototyping and Flying Components

De Havilland embarked on a programme to build a flying component and embark on large sub-component testing. In mid-1988 a DASH-8 inboard flap, see Figure 12, was chosen because it met size constraints, see previous discussion on autoclave development, and it was a nearly flat panel. As it was, the 96" long panel still had to be manhandled in an out of the spray booth to apply the BR-127 adhesive primer. This component first flew on April 25, 1991 as the largest commercial FML structural component.[2] A year later the first production C-17 cargo door flew. It was larger and was slightly double-curved and thus new production processes had to be addressed, a slight stretch-form and spliced joints were applied. Details of these techniques, using recommendations of bonded joints from the PABST [37] programme would be disclosed by Pettit [38, 39]. At this time, the term *fibre metal laminate* was coined.

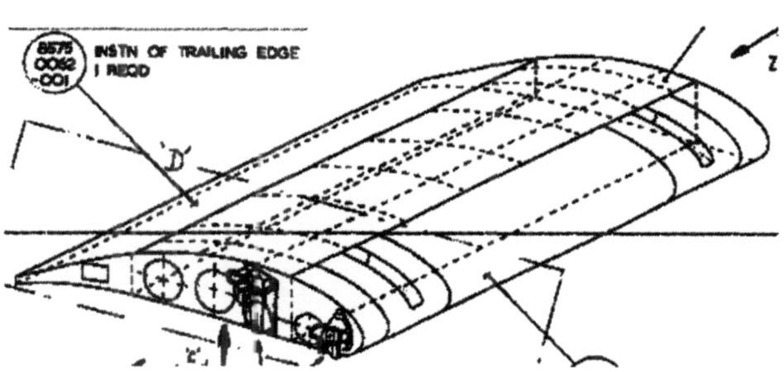

Figure 12: DASH-8 I/B flap.

[1] The last design with extensive primary structure metal-to-metal bonding still in production is the DASH-8 series of aircraft.

[2] The F-50 prototype with Arall lower wing inspection covers flew in October 1987.

Panel size difficulties encountered on the flap development all but disappeared on DADT barrel testing on fuselage panels with bonded body stringers in 1990 where 48" x 12" panels were readily available. Testing on Arall 3 panels with fibre in the hoop direction and developmental versions of Glare 3 proved very successful [40, 41]. Weight savings and part consolidation were attractive features. A major limitation, much as in early days of large plate development, see remarks by Clotworthy [13], was that the panels were too small. In the preliminary design phase on the DASH-8 series fuselage in the early 1990s, results suggested a requirement of 8 circumferential panels and 5 panels end to end. Thus the weight of the joints detracted up to 80% of the weight savings, let alone the adverse effect on the fabrication costs of the fuselage. Work by Garesche [42] and colleagues, along with work of Pettit [38], paved the way to wide spliced laminate development, to the point where fibre metal laminate fuselage panels wider than current aluminium alloy sheets limitations (about ~110" wide for a 0.125" thickness), can now be made as thin as 0.032". Material handling challenges of such a panel are easily overcome. An 87" x 144" DADT spliced panel is shown in Figure 13, one of two concept designs scheduled for testing this year (ed.: 2001). Two 124" x 156" panels, see Figure 14, are to be incorporated on the S400 static test article for static testing, see Figure 15, towards the end of this year (ed.: 2001).

Figure 13: Wide spliced FML DADT panel.

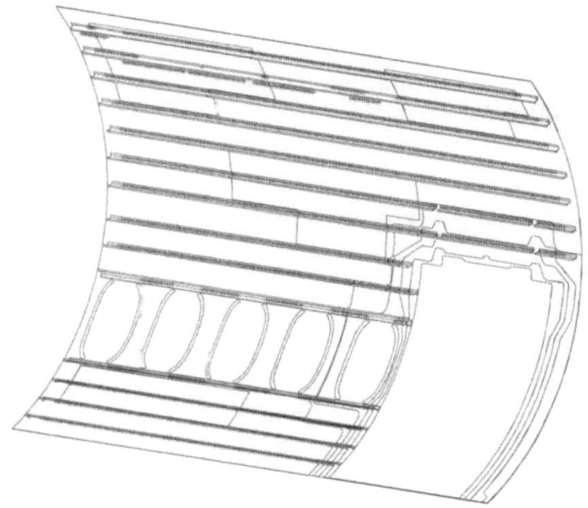

Figure 14: Stbd. CAST panel design.

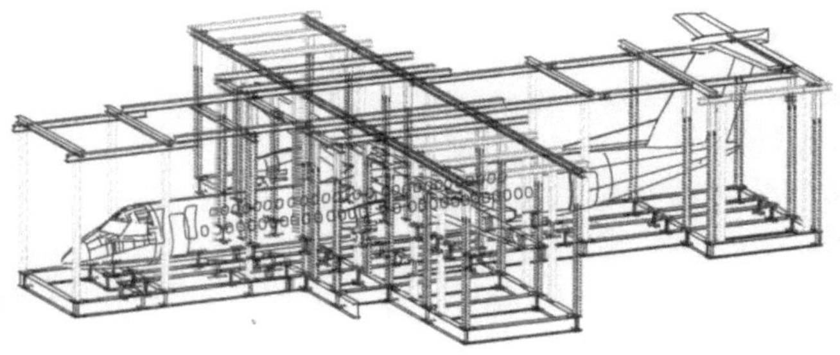

Figure 15: DASH-8 series 400 CAST fixture.

Figure 16: NACA I-25 series type shear web.

Further work with National Research Council of Canada / Institute for Aerospace Research (NRC/IAR) [43] will focus on post-buckled strength analysis of flat and curved panels, see Figure 16, and thus revisiting work of some 50 years ago by Clark [44], Kuhn [45] and Peterson [46].

The future of fibre/metal laminates

Nearly 30 years after the initial development fibre metal laminate components flew on an aircraft, the A380 will go aloft with some 1.5 t of the material in the fuselage. To achieve this, the design and manufacturing difficulties that will inevitably arise will be resolved. To look ahead, one can postulate higher temperature versions [47, 48] of the materials will need to be developed to support 'Sonic Cruisers' and super-cruise aircraft. The drive for yet larger panels may see the re-introduction of in-situ bonding on the production lines, as on the Comet. It is unlikely that resistive heating will be used and laser bonding could be the way forward. To be sure, developers of the technology in the future will no doubt look back to garner from lessons learned from the past.

References

[1] John Riley, Ed., Alcoa Technology Report to the Aerospace Industry, Vol. 8, February 1989.

[2] Jack McKillop, http://www.microworks.net/pacific/aviation/ rb_conestoga.htm.

[3] 'Plastics at Hatfield', in: *de Havilland Gazette*, No. 22, Feb. 1939: 4-5.

[4] C.J. Moss, 'Redux Bonding of Aircraft Structures', in: *Journal of the Royal Aeronautical Society*, Vol. 54, 1950: 640-650.

[5] R.E. Bishop, 'The Comet as a Design Project', in: *de Havilland Gazette*, No. 69, June 1952.

[6] H. Povey, 'Planning and Production Methods used in the Construction of the de Havilland Comet', in: *Journal of the Royal Aeronautical Society*, Vol. 55, August 1951.

[7] L.J. Hart-Smith, 'Adhesive Bonding of Aircraft Primary Structures', presented to SAE Aerospace Congress & Exposition, Los Angeles CA, October 13-16, 1980.

[8] Edw. J. van Beek, 'Design Aspects of Bonded Structures: Use of Redux in the Fokker F-27 Friendship', in: *FLIGHT*, October 25, 1957.

[9] F.H. Parker, 'Metal Adhesive Processes', in: *Journal of the Royal Aeronautical Society*, Vol. 55, March 1951: 153-168.

[10] David W. Jensen, *ARALL Short-Beam Compression and Shear Panel Buckling*, Alcoa Contract No. CE 362682, December 12, 1989.

[11] L. Ross Levin, *Ultimate Stresses Developed by 24S-T and Alclad 75S-T Aluminum-Alloy Sheet in Incomplete Diagonal Tension*, NACA TN-1756, November 1948.

[12] L. Ross Levin and David H. Nelson, *Effect of Rivet or Bolt Holes on the Ultimate Strength Developed by 24S-T and Alclad 75S-T Sheet in Incomplete Diagonal Tension*, NACA TN-1177, January 1947.

[13] H.J. Pollard, 'New Materials and Methods for Aircraft Construction', in: *Journal of the Royal Aeronautical Society*, Vol. 57, May 1953: 277-300.

[14] Paul Badre, 'Modern Methods of Aircraft Production', in: *Journal of the Royal Aeronautical Society*, Vol. 61, June 1957.

[15] 'Arrow: A World-Leading Interceptor by Avro Aircraft', in: *FLIGHT*, October 25, 1957.

[16] *The Aeroplane*, March 8, 1957: 322.

[17] S.W. Allen and T.R. Truax, *Glues Used in Airplane Parts*, NACA Report 66, 1920.

[18] John Black and R.F. Blomquist, *Relationship of Metal Surfaces to Heat-Aging Properties of Adhesive Bonds*, NACA TN-4287, September 1958.

[19] John Black and R.F. Blomquist, *Development of Metal-Bonding Adhesive FPL-710 with Improved Heat-Resistant Properties*, NACA RM-52F19, July 8, 1952.

[20] John Black and R.F. Blomquist, *Development of Metal-Bonding Adhesives with Improved Heat-Resistant Properties*, NACA RM-54D01, May 14, 1954.

[21] H.W. Eickner, W.Z. Olson and R.F. Blomquist, *Effect of Temperatures from -70 °F to 600 °F on Strength of Adhesive-Bonded Lap Shear Specimens of Clad 24S-T3 Aluminum Alloy and of Cotton and Glass-Fabric Plastic Laminates*, NACA TN-2717, June 1952.

[22] J.G. Kaufman, *J. Basic Eng.*, 89, 1967: 503-507.

[23] J.G. Kaufman, *Fracture Toughness of 7075-T6 and -T651 Sheet, Plate, and Multilayered Adhesive-Bonded Panels*, ASME PAPER 67-MET, Apr. 1, 1967.

[24] W.S. Johnson, W.C. Rister and T. Spamer, 'Spectrum Crack Growth in Adhesively Bonded Structure', in: *J. of Eng. Matl. Tech.*, ASME, Vol. 100, 1978: 57-63.

[25] W.S. Johnson and J.M Stratton, 'Effective Remote Stresses and Stress Intensity Factors for an Adhesive Bonded Multi-ply Laminate', in: *Eng. Frac. Mechanics*, Vol. 9, 1977: 411-421.

[26] W.S. Johnson, 'Damage Tolerance Evaluation of Adhesively Laminated Titanium', in: *J. of Engineering Materials and Technology*, ASME, 105, 1983: 182-187.

[27] J. Schijve, H.T.M. van Lipzig, G.F.J.A van Gestel and A.H.W. Hoeymakers, 'Fatigue Properties of Adhesive-Bonded Laminated Sheet Material of Aluminum Alloys', in: *Eng. Fracture Mechanics*, Vol. 112, 1979: 561-579.

[28] H.T.M. van Lipzig, *The Retardation of Fatigue Cracks*, Thesis Delft University of Technology, 1973 (in Dutch).

[29] G.F.J.A. van Gestel, *Crack Growth in Laminate Sheet Material and in Panels with Bonded Integral Stiffeners*, Thesis Delft University of Technology, 1975 (in Dutch).

[30] A.H.W. Hoeymakers, *Fatigue in Lugs*, Thesis Delft University of Technology, 1975 (in Dutch).

[31] Jacobus Schijve, Boud Vogelesang and Roelof Marissen, *Laminate of Aluminum Sheet Material and Aramid Fibers*, US Patent 4,500,589, Feb. 19, 1985, filed Sep. 20, 1983. Dutch patent # 8,100,087, Jan. 9, 1981.

[32] M.L.C.E. Verbruggen, *Aramid Reinforced Aluminium Laminates: ARALL*, Report LR-509, PhD Thesis Delft University of Technology, 1986.

[33] S.G. Lekhnitskii, *Anisotropic Plates*, translated S.W. Tsai and T. Cheron, New York NY: Gorden & Breach, 1968 (originally published in Russian as a monograph 1944).

[34] Paul Paris and George Sih, *Stress Analysis of Cracks*, ASTM STP381, 1965: 30-81.

[35] L.H. van Veggel, A.A. Jongebreur and J.W. Gunnink, 'Damage Tolerance Aspects of an Experimental ARALL F-27 Lower Wing Skin Panel', in: Proceedings of the 14th Symposium of the International Committee on Aeronautical Fatigue, June 8-12, 1987, Ottawa, Canada: 465-502.

[36] M. Ioannou, L.J.J. Kok, T.M. Fielding and N.J. McNeill, 'Evaluation of New Materials in the Design of Aircraft Structures', in: Proceedings of the 14th Symposium of the International Committee on Aeronautical Fatigue, June 8-12, 1987, Ottawa, Canada: 127-149.

[37] L.J. Hart-Smith, 'Difference Between Adhesive Behavior in Test Coupons and Structural', presented to American Society for Testing and Materials Adhesive Committee, Phoenix AZ, March 11-13, 1981.

[38] Richard Pettit, *Fiber/Metal Laminate*, US patent 5,227,216, July 13, 1993, filed March 25, 1991.

[39] Richard Pettit, *Fiber/Metal Laminate Splice*, US Patent 5,951,800, Sep. 14, 1999, filed Nov. 26, 1997.

[40] Vogelesang and Roebroeks, *Metal -Resin Laminate Reinforced with S2-Glass Fibres*, US patent 5,039,571, Aug. 13, 1991, filed Oct. 11, 1988.

[41] Roebroeks and Mattousch, *Impact Resistant Laminate*, US patent 5,547,735, Aug. 20, 1996, filed Oct. 26, 1994.

[42] Carl Garesche, Gerardus Roebroeks, B. Greidanus, Rob van Oost and Jan Willem Gunnink, *Spliced Laminate for Aircraft*, US Patent 5,429,326, July 4, 1996, filed July 9, 1992.

[43] Alexandre Jodion, G. Shi, C. Poon and P. Straznicky, *Experimental Analysis of Diagonal Tension in GLARE*, CASI 14th Structures and Materials Symposium, April 29 - May 2, 2001, Toronto CA.

[44] J.W. Clark and R.L. Moore, *Torsion Tests of Aluminum-Alloy Stiffened Circular Cylinders*, NACA TN-2821, November 1952.

[45] Paul Kuhn, James P. Peterson and L. Ross Levin, *A Summary of Diagonal Tension PART II - Experimental Evidence*, NACA TN-2662, May 1952.

[46] James P. Peterson, *Experimental Investigation of Stiffened Circular Cylinders Subjected to Combined Torsion*, NACA TN-2188, September 1952.

[47] W.S. Johnson, Ted Q. Cobb, Sharon Lowther and T.L. St.Clair, *Hybrid Titanium Composite Laminates: A New Aerospace Material*, 21st Annual Adhesion Society Meeting, Savannah GA, February 22-25, 1998.

[48] J. Cook, 'Properties and Processing of Novel High Temperature Laminates', presented at Aeromat '92, May 20, 1992, Anaheim CA.

Response 4

Fedde Holwerda

National Aerospace Laboratory NLR

Up till now, Mr. Schatzberg and the respondents have reviewed ninety-eight years of aircraft design and development. I will concentrate on the thirty or so years of my own experience, however. When I started at Fokker, I was behind a drawing board, where one of the first lessons you learn, having left the ideal world of the Faculty of Aerospace Engineering at Delft, is that designing an aeroplane is a continuous process of finding the best compromise.

This may not be such a good start to my lecture, since compromise is a bit of a dirty word in the science and engineering worlds. I fully agree with Mr. Campbell who made this quite clear. In my view, the motto for today could be: 'Hit for the limits but go for the sellable compromise', where the best compromise is what we would term the 'optimum design'.

So where are we today regarding structural designs and materials applications as far as the designer is concerned? In terms of structural efficiency, safety and durability the designs are already mature. We are already a long way down the learning curve when using high-strength alloys, and only incremental improvements with numerous trade-offs can still be made with these materials. These may take many years to be validated, leading to higher cost, both for materials and during certification.

New materials require new design approaches, entailing development risk that can be tackled only by time-consuming validation programmes. Experience, together with a healthy dose of scepticism, shows us that the advocates of improvements through new materials are always over-optimistic. The aerospace industry is actually quite conservative – new materials are adopted far quicker for application in golf clubs, racing cars, racing boats and even in pleasure-boats. This is mostly due to the fact that the qualification and certification process in the aviation industry is so lengthy.

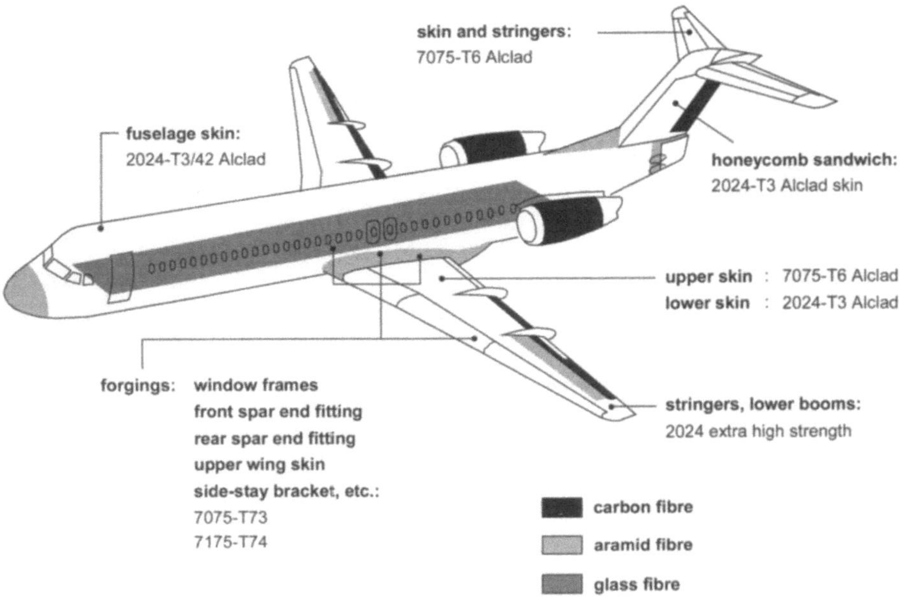

Figure 1: Aluminium and composite applications in a modern civil transport aircraft (Fokker 100).

If we look at a modern civil transport aircraft designed in the nineteen-eighties, in this case the Fokker 100, we can see what was state-of-the-art for that time, see Figure 1. The airframe is 65% aluminium, 10% composites and 25% miscellaneous. If we compare this with today's regional airliners from Bombardier, Fairchild and Embraer, we find a small increase in the application of composites but otherwise the design concept differs very little.

If we look at Airbus and the A380, and the new challenges tackled by this torch-bearing design for the state-of-the-art circa 2005, then we see real breakthroughs, i.e. composites in primary structures such as the tail and the centre wing box and a fuselage incorporating welded stringers and Glare, see Figure 2. The great technical challenges that need to be overcome in order to meet the requirement goals of this aircraft make these innovations necessary.

Let us go back to the Fokker 100 and the application of Glare now. About ten years ago, when I was chief of engineering during the design of the Fokker 70, I was visited by Daan Krook and Jan Willem Gunnink. They tried to convince me that not introducing Glare as the fuselage material would be the greatest mistake of my life. Their arguments were that its lightness and better durability, as well as Fokker's bonding experience, would make it a great success. However, I was not looking for weight or durability gains over the Fokker 100 and Fokker 28, since these aircraft were already the world benchmark. I was looking for cost and risk reduction on the lead time instead. After the Fokker 100 I had only two targets, i.e. 'on time' and 'within costs', and unfortunately Daan and Jan Willem were not bringing me solutions for these with Glare. It would take another ten years before Glare would be ready.

Another example of learning curve maturity can be found with the improved 7000-series aluminium alloys. These materials, developed in the 1940s, had seen several attempts at improvement through the years. Figure 3 shows the attempts in the 1970s and 1980s to achieve 15% weight savings when compared to the industry standard 7075 alloy. Considerable metallurgical, manufacturing and test programme knowledge was required to achieve only incremental improvements. This situation has not changed, so fundamental breakthroughs in this field cannot be expected. There is no better example to underline Mr. Schatzberg's statement that in hindsight the change from wood to aluminium was a direct hit; both the 2024 and the 7075 alloys proved unbeatable for many years.

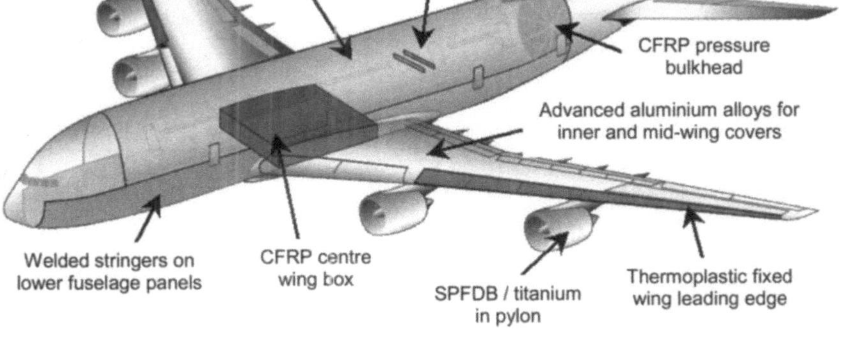

Figure 2: New aerospace materials applications in a modern civil transport aircraft (Airbus A380) [1].

There are other examples of new materials that make new fatigue design approaches necessary, such as inter-metallics and ceramics. In Figure 4 we see a compendium of crack growth rates of three classes of materials, i.e. conventional metal alloys, inter-metallics, and ceramics and ceramic composites. The conventional metal alloys can be considered for both damage-tolerant and safe-life fatigue design categories including fatigue crack growth. The inter-metallics have reasonable threshold ΔK-values but their crack growth curves are steep. Therefore, crack growth cannot be permitted and damage-tolerant design would have to rely on ΔK-threshold. Nevertheless, a mixed design practice using ΔK-threshold and S-N fatigue limit is likely to be the best option. The ceramics and ceramic composites have very low threshold ΔK-values, and their very steep crack growth curves make them more suited for safe-life designs. However, these options are likely to be in the distant future, when heat-resistant structures with high durability may be necessary. The introduction of new materials is therefore a process that takes decades if it is to be achieved at all.

New design approaches also lead to new certification requirements for any new materials applied. We have seen that with Glare the certification process alone took many years. I experienced such a process myself at Fokker Engineering some fifteen years ago, when we changed, together with DASA and Shorts, the Redux bonding process from a liquid/powder system to a Redux film. This was no great change, and even though the manufacturing process tolerances were much smaller than those required, the validation process was still exhaustive. We are going through the same tough and costly process with Glare right now.

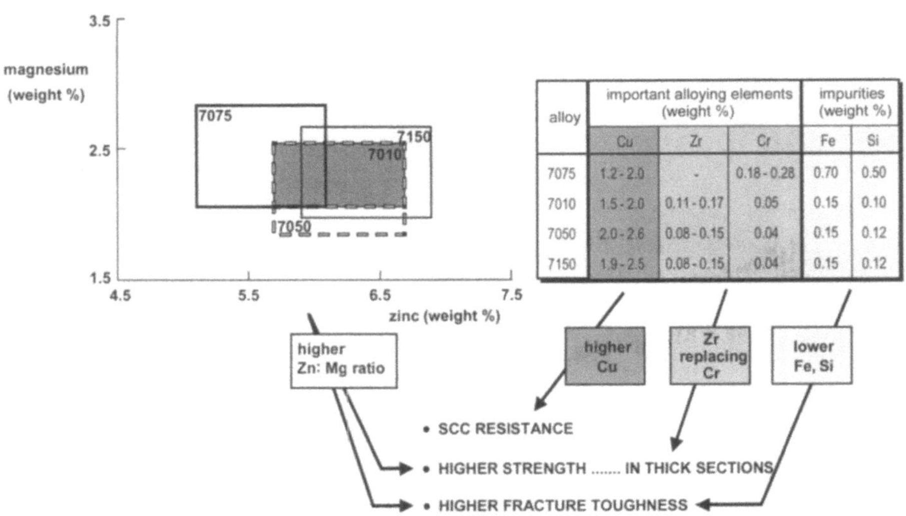

alloy	important alloying elements (weight %)			impurities (weight %)	
	Cu	Zr	Cr	Fe	Si
7075	1.2 - 2.0	-	0.18 - 0.28	0.70	0.50
7010	1.5 - 2.0	0.11 - 0.17	0.05	0.15	0.10
7050	2.0 - 2.6	0.08 - 0.15	0.04	0.15	0.12
7150	1.9 - 2.5	0.08 - 0.15	0.04	0.15	0.12

Figure 3: Example of learning curve maturity; improved 7000-series (Al-Zn-Mg-Cu) aluminium alloys.

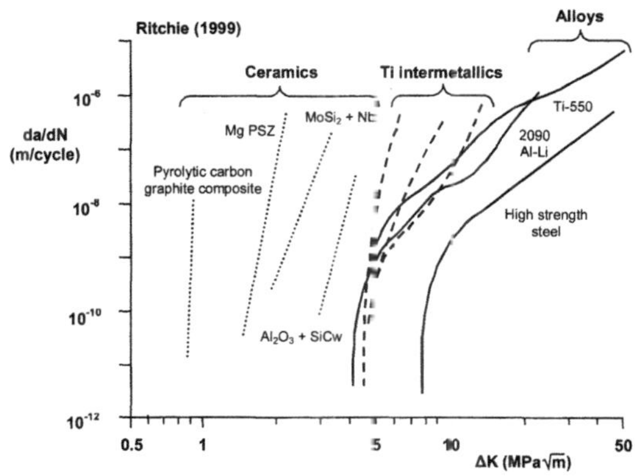

	Fatigue design categories		
	Damage tolerance		Safe life
	Crack growth	ΔK-threshold	
Alloys	✓	✓	✓
Ti inter-metallics	✗	✓	Linked ✓
Ceramics	✗	✗	???

Figure 4: New fatigue design approaches for new materials (inter-metallics, ceramics) [2].

There is a well-known law of constant energy, yet I would argue that there is also a law of constant troubles. Take the damage tolerance approach for example, i.e. safety by inspection and repair when damage is found. In this respect, Glare would seem to be an ideal material, i.e. hardly any crack growth and hardly any loss of residual strength, see Figure 5. Yet, as a certain Dutch soccer player and philosopher says: 'Every advantage has its disadvantage.' For Glare there are more small cracks, which we the designers say are harmless, but the certification authority wants to be able to exclude any doubts. More importantly, we need to eliminate the doubts of our customers, since, as Kees de Koning says: 'Who wants to finance twenty years of research with an undefined outcome?'

In conclusion, the development of a new material for widespread application is a tough process – a road travelled with sweat and tears. Of all the different fibre metal laminate research projects carried out since the nineteen-seventies we can now say that Glare made it. Maybe Glare is therefore what Mr. Campbell described as 'unobtainum'. I would describe it as a world-record breaking accomplishment for all those who have kept their belief in the material and worked so hard for so many years for it. I would say that this is typically Dutch; maybe such 'unobtainum' can only be discovered here in the Netherlands! I always tell foreigners that it would not even have been possible to live here in the Netherlands without Dutch innovation, as has been proven several times in the past, and maybe once again with our invention of 'unobtainum'. The application of Glare in the A380 will be the reward for all those who persevered, not least Boud Vogelesang on the brink of his retirement. Boud has overcome many obstacles and has made it and I congratulate him and his team wholeheartedly. Nevertheless, the work on Glare is not finished; there are still other promising fibre and metal combinations to do research on – and that will continue, just as we are still working on aluminium alloys, even though they were introduced before the Second World War as I showed earlier. It is our business and we love it!

References

[1] EADS Airbus GmbH.

[2] R.O. Ritchie, in: *Engineering against Fatigue*, eds. J.H. Beynon, M.W. Brown and R.A. Smith, Rotterdam: A.A. Balkema Publishers, 1999.

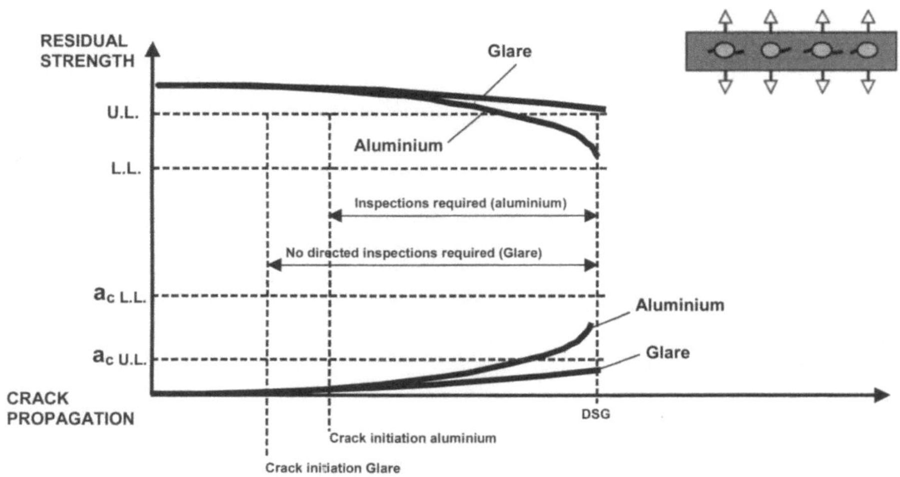

Figure 5: Application of damage tolerance philosophy to a Glare fatigue case (not accidental damage).

Response 5

Karl-Heinz Rendigs

Airbus Deutschland GmbH

It is an honour for me to tell you something about the work we have carried out during the last ten or fifteen years to improve the aluminium structures at Airbus. I will quickly cover the status of aluminium structures today, then give some ideas of development work mainly on new fuselage structures using laser beam welding, friction stir welding, large extrusion and large cast components and finally give some words about Glare.

A brief history

If we look at the material distribution in Airbus aircraft, we see that around 80% of the Airbus A300 was made from aluminium and only about 5% from composites. At the other end of the scale, in the A380 we see a large decrease in aluminium application and a significant increase in the amount of carbon fibre. The amount of titanium stays roughly the same and there is a reduction in the use of steel.

Let us consider the A320 in more detail. The wing and fuselage were made from aluminium. The vertical and horizontal stabilisers and other moving parts on the wing were made from carbon fibre, the landing gear from steel and the pylons from titanium. The skin of the fuselage was made from 2024 clad sheet and the stringers were also made from this material, either as sheets or extrusions. The frames were made from 2024 and high-zinc alloys. Seat rails were 7175 extrusions, floor beams were high-zinc 7000 alloy extrusions and window frames were 7175 forgings. The wing slats were made from a special high-temperature 2618 alloy, the D-nose was made from 2024 sheets, the top wing panel was 7150, the bottom panel was 2024. The flap supports were forgings, while the flap tracks were precision castings. The landing gear supports were forgings.

We have also developed a variety of very complicated aluminium precision castings over the last ten or fifteen years. Many thousands of precision-cast tracks are currently flying successfully in Airbus aircraft, however there were many hurdles to be crossed during their development, especially where stress was concerned. These days, however, they are routine products.

New methods for the A380

The A380 will be very large, i.e. if we compare it to the A319, we find that the horizontal tailplanes of the A380 are as large as the A319's entire wing! This brings a number of new questions about how such a large aircraft can be constructed. Ten years ago, we started to consider several ideas regarding the riveted structure of the fuselage. The differential design is currently nearly at the limit of possible automation, with many large machines carrying out the riveting of most modern aircraft. We therefore wanted new ideas to move from a differential design to an integral one. These included for example using extruded stringer/skin connections, brazed connections and laser welding. A few years ago we dropped the brazing project since this method turned out not to be so economical. We continued with the extruded and laser welding alternatives, however, and at the moment it seems that laser welding holds the most promise.

The main goals for these integral structures are cost reductions of 15% and weight reductions of 10%. It also turns out that the removal of the rivet holes leads to large improvements in the corrosion aspects as corrosive initiation points are removed from the structure.

Laser beam welding

The advantages of laser beam welding are that it delivers a very narrow, deep seam, there is a low input of energy into the aluminium, it provides a high welding speed and it takes place in a normal atmosphere. The stringer is welded to a sock on the inside of the skin. The sock is created by chemical milling and is there so that the rest of the skin remains unaffected by the welding process, i.e. otherwise there would be problems with changes in properties across the skin section.

Two lasers are used to weld the skin to the stringer and this has being trialed since 1997 using equipment at the Nordenham plant. These lasers are fitted to the top of an eight-metre high gantry and pass through a special head that moves over the fixed material that is to be welded. The resulting microstructure is very homogeneous.

After much qualification work, we successfully produced the first laser-welded production shell in November 2001, and we are proud to have been the first to achieve this. We are currently busy with qualification for the A380.

Friction stir welding

A new process with great potential is friction stir welding. We are currently at the same level with this process as we were with laser welding about six years ago. There may be an opportunity in the future to change the longitudinal and circumferential joints to friction stir welds if this turns out to be beneficial. The process is very simple, i.e. a pin and a good milling machine can cause two materials to fuse, maintaining 90% of the normal material properties.

There are remarkable advantages to using this process. For one thing, the temperatures remain low, i.e. 400-500 °C. We can also join all types of different aluminium alloys together. There are no cracks, pores, noise, smoke or shield gas. The process produces a wrought microstructure, high mechanical properties and a minimum of distortion. There seem to be many advantages and the only disadvantage is that you need a very good clamping system to hold both parties together during the process.

Integrated skin and stringers

We investigated the possibilities of extruding an integral skin and stringer assembly in one pass. Eventually we chose a tube with external stringers, which is then cut open and formed to the curvature of the fuselage. This research was carried out in conjunction with Russians in Moscow and Samara, where they have the largest press in the world, which allows us to reach large maximum diameters. The tests are still going on and it is still too early for the application of this technique. At the moment, we are running fatigue tests on large components made using this method.

Castings

We have also carried out tests to produce much larger castings, such as a single-cast baggage compartment doorframe. We have also managed to produce sand-cast integral passenger doors with our partners in the US. This incorporates a complete skin, which is later milled and polished. While this technique is nearly ready for application, it will not be used for the production of the A380, because the risk was deemed to high as a result of price increases and the departure of several key figures at the foundry where the methods were developed. It is our intention to use such techniques for large complicated single-castings in the future as a way to save money.

Materials

We will be using Al-Mg-Si-Cu 6000-series alloys for our welded sections, but we also plan to use new alloys in the future such as the Al-Mg-Sc and Al-Mg-Li alloys. The major advantages of 6013-T6 over 2024-T3 are that it is weldable, highly formable in a T4 condition and is not very expensive. The 1424-TX lithium alloys developed together with the Russians have the highest potential for weight savings. Al-Mg-Sc proves to be the best corrosion-resistant, weldable high-strength aluminium alloy and has a lower weight.

The main driving forces for Glare were the potential weight savings, fire resistance and tailored properties within the structure. At the moment we are planning to use Glare in the upper fuselage and crown sections, and we are also discussing planned changes from aluminium to Glare in some other areas.

Summary

We have introduced completely new technologies for the Airbus A380. It also has a completely new geometry and we hope that this will make its introduction in 2006 successful and of benefit to the market.

Day 3: NEW MATERIALS AND SAFETY

Sponsored by: AIRBUS

Keynote lecture

The material down-selection process for A3XX

Jens Hinrichsen

Airbus Industrie
Large Aircraft Division

Abstract – This paper illustrates the technology selection process for a new long-range aircraft family. These future members of Airbus Industrie's product family will continue the evolution of advanced technologies at Airbus and will also pioneer new technologies. Guiding principles and the elements of the down-selection process will be presented. The close link between structural design criteria, material properties and manufacturing processes will be outlined for two different examples, i.e. selection of Glare application for fuselage panels and a discussion of alternative manufacturing processes for non-pressurised fuselage sections in CFRP.

Key words – Airbus, A3XX, technology selection, aircraft structures, materials, manufacturing processes.

Goals of the down-selection process

The goals of the down-selection process are:

- Deliver a robust structural design and mature material and manufacturing technologies.
- Prepare technical solutions that allow the achievement of target weights and target costs for the airframe.
- Keep advanced technologies within the technology evolution at Airbus Industrie.
- Mitigate risks from initial steps into new technologies.
- Support standardisation of structural design and maintenance concepts across all Airbus programmes.

In order to achieve these goals, guiding principles have been established. The elements of the down-selection process follow these principles and deliver all information on technical aspects. Final decisions are taken at the aircraft programme level, targeting a well-balanced product definition that reviews business issues and trade studies for manufacturing costs versus weight savings for all proposed technical solutions. Materials and manufacturing process down-selection supports this programme management decision process.

Principles of the down-selection process

The principles of the down-selection process are:

1. Transfer 'the lessons learned' from the existing Airbus fleet into A3XX concepts.
2. Continue technology evolution at Airbus Industrie.
3. Take benefit from earlier Airbus programmes.
4. Prepare 'Right First Time' for series production.
5. Establish targets for trends of technology parameters versus time.

Principle: Transfer 'the lessons learned' from the existing Airbus fleet into A3XX concepts

Before entering into any discussion of advanced or new materials, we have to assure that 'the lessons learned' from operations of the existing Airbus fleets are incorporated into the structural design concepts of a new product. Through workshops with maintenance experts from key airlines, 'in-service experience' concerning topics like corrosion protection and sealant, inspections and repairs for crack growth as well as accessibility and tooling are revisited. Results from such collaborative work enter the Structural Requirements Document (StRD), which is the basis for design work for each of the major aircraft components. This very basic principle follows the Airbus Industrie policy at the corporate level, which aims for continuous product improvement and 'design for maintainability'.

Principle: Continue technology evolution at Airbus Industrie

Inaugural use of a new technology shall proceed step by step, building on experience with earlier Airbus products, as shown in Figure 1 for composites. The background for such limitation stems from the 'teething troubles' experienced with almost every new technology. The reasons are many. Firstly, the stability of production processes at the shop-floor level has to be achieved, where theoretical simulation or production trials under lab conditions may fail to pave the way sufficiently. Then, maintenance staff at airlines have to go through their learning curve within the tough operational environment of an aircraft, which is characterised by external damage from hailstones, lightning strikes, birdstrikes, debris from taxiways and runways, also trucks running into primary structure during ground-service, etc. Furthermore, the structure to be maintained is subjected to aggressive fluids and to temperatures changing between -60 °C and +110 °C. The prediction of structural behaviour and maintainability through the whole aircraft life – about 25 years – is limited. There are 'broken bones' in the industry all around the world, resulting from applications of new technologies, i.e. water ingress with Aramid fibres and de-bonded longitudinal lap joints of metal fuselage skins, to mention just two prominent issues. These examples already illustrate that the behaviour of structures in service depends not only on material performance but also on design solutions and manufacturing capabilities. A learning process has to be established for new technologies, one that allows for the optimisation of materials and processes, and increasing areas of application versus time. Figure 2 displays evolutionary steps in the learning curve for material applications over almost 30 years and throughout the full range of Airbus products, leading up to the A3XX.

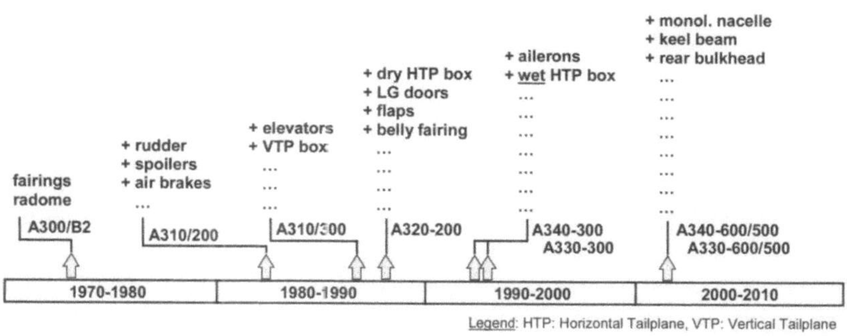

Figure 1: Evolution of composite material applications at Airbus.

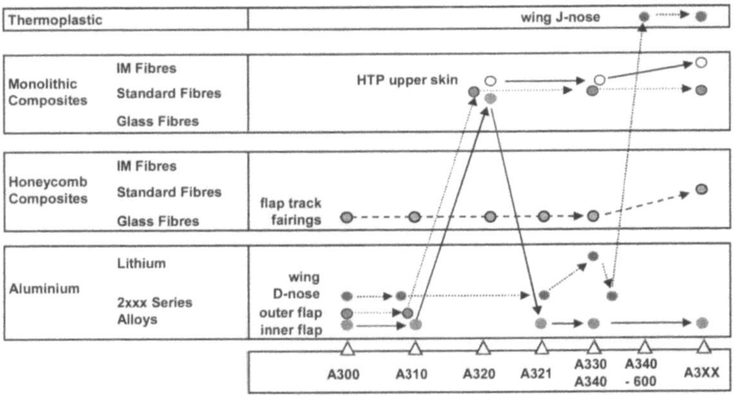

Figure 2: Product improvement evolution leading up to A3XX.

Principle: Take benefit from earlier Airbus programmes

The introduction of new technologies to Airbus aircraft that will be certified earlier than the A3XX supports the risk mitigation plan for the new product family. Development test results are available, qualification of materials and manufacturing processes is completed and series production established.

Achievements such as these must be combined with dedicated tests that principally address the requirements specific to the A3XX, namely those being linked to the unique size of components and differences in design concepts, static loading and complexity of surface curvature.

These are the main parameters driving the down-selection of materials and manufacturing technologies. Knowledge of the constraints of a new technology is the biggest advantage taken from earlier programmes – the aim is *not* to copy exactly. Figure 3 displays major new technologies, which will have been proven through earlier Airbus programmes. For these, the A3XX is second in the chain of evolution.

Principle: Prepare 'Right First Time' for series production

The top target for a series production must be 'Right First Time'. To come as close as possible to this target requires simulation of manufacturing processes in a plant environment – not in laboratories. The test articles should be of equivalent size and surface curvature, see for example Figure 4. Also, stiffening elements and local reinforcements at load introductions should be demonstrated in tooling and manufacturing processes, representing a real structure at full scale – not a generic structure.

For example, performing hand-layup of CFRP preforms on a mould of real size gives a feeling for the accessibility and quality that can be achieved in a production line. Development engineers get information about gaps and overlaps of the layup, which need to be in line with structural design requirements. Different inspection methods can be studied in order to optimise the manufacturing processes and the quality control efforts. Subsequent modifications of jigs and tools can be performed before the production line starts operations. The resources spent on manufacturing demonstrators will pay off; rework in production lines and the risk to fix problems on aircraft in service can be minimised. Last but not least, a demonstrator creates motivation for those people doing the job in future.

Principle: Establish targets for trends of technology parameters versus time

It is one of the most challenging tasks in aircraft development to achieve maturity of materials and manufacturing technologies in time for the programme launch. The programme schedule in Figure 5 shows that the freeze of concepts for design and manufacturing has to take place ahead of programme launch. Following milestone 'ATO', guarantees for aircraft performance and prices are negotiated with airlines. At this point in time, airframe weight and cost need to be known within small margins, and this is strongly linked to major technologies. All activities on the right-hand side of this key milestone are dedicated to detailed product definition as well as preparation of series production. This also includes orders for materials, tools and equipment.

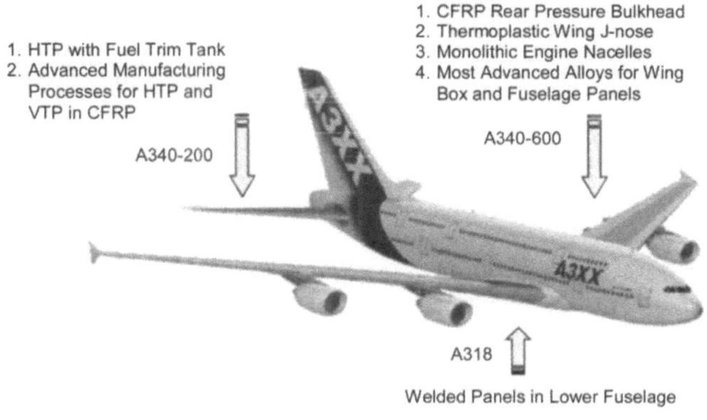

Figure 3: A3XX benefits from earlier Airbus programmes.

Figure 4: Bulkhead manufacturing demonstrator (left) and Glare panel installation on aircraft (right).

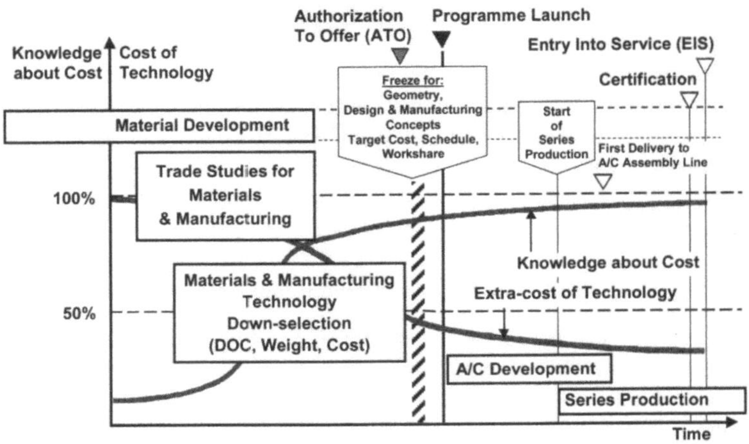

Figure 5: Trends of technology parameters versus time.

Technology preparation takes place to the left of the programme launch milestone and starts at the far left of the time schedule. At this point in time the *Knowledge about Cost* of a new technology is at a poor level and the *Extra-cost of Technology* is likely to be overestimated, see Figure 5. It is now recommended by the proposed principle that targets are settled for these two driving parameters:

1. Cut *Extra-cost of Technology* by a factor of two (for the example in Figure 5).
2. Be prepared to determine cost within a scatter of 5% at decision milestone; this is equivalent to *Knowledge about Cost* at a level of 95%.

To set such bold targets at the very beginning creates 'the right mindset' for all tasks to be performed for the down-selection.

Elements of the down-selection process

Apart from ongoing work in the field of materials and process development, initial preparations for the A3XX started in 1994, as shown in Figure 6. There was a logical sequence of milestones for goals and of tasks when preparing for the A3XX. An outline of the down-selection process needs simplifications: the various activities are clustered in so-called elements of the down-selection process.

These elements will be described in the following paragraphs and it will become evident that they are very different in nature. The reader will find product strategy and policy, structural design drivers and materials performance, production cost estimates/targets and comparison of manufacturing processes all linked to each other.

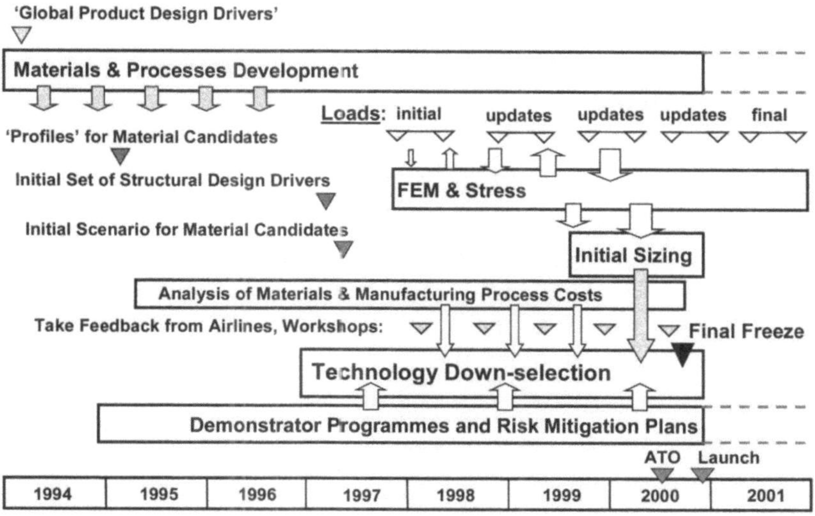

Figure 6: Elements and tasks of the down-selection process versus time.

Element: Identification of 'Global Product Design Drivers'

This activity took place in spring 1994, at first delivering scenarios for future projects versus time as shown in Figure 7. A horizontal beam for each project starts at the assumed programme launch and runs out at entry into service (EIS). Now, 6 years later, we know better. The 'New 250-seater Twin', a successor to the A310, has not yet been realised. 'A340 advanced' is scheduled to enter service as the A340-600 in 2002, one year later than in the scenario. The 'FLA' project is now known as A400M, a military transport aircraft for which the concept phase is almost complete. The basic variant of A3XX was launched on December 19, 2000, finally named A380-800, two years later than expected in early 1994.

Anticipated projects and schedules of such scenarios will always fail to some extent to predict the future. Nevertheless, it can be regarded as a simple and powerful tool, helping to identify *target projects* and the time frame for technology preparations. The *drivers for technology programmes* are derived from a set of 'Global Product Design Drivers' for each of the projects. Five categories have been established, i.e. *Improvements*, *High Performance*, *Low First Cost*, *Low Cost of Ownership* and *Robust Design*. Each project is then characterised by giving a first priority to one of these categories, indicated by a full circle in the relevant column of Figure 7. A second priority is also given, marked by a white filled circle. These priorities highlight the competitive edge of the future aircraft product in the market place. It is a 'must' to achieve this competitive edge, which is measured against the competition.

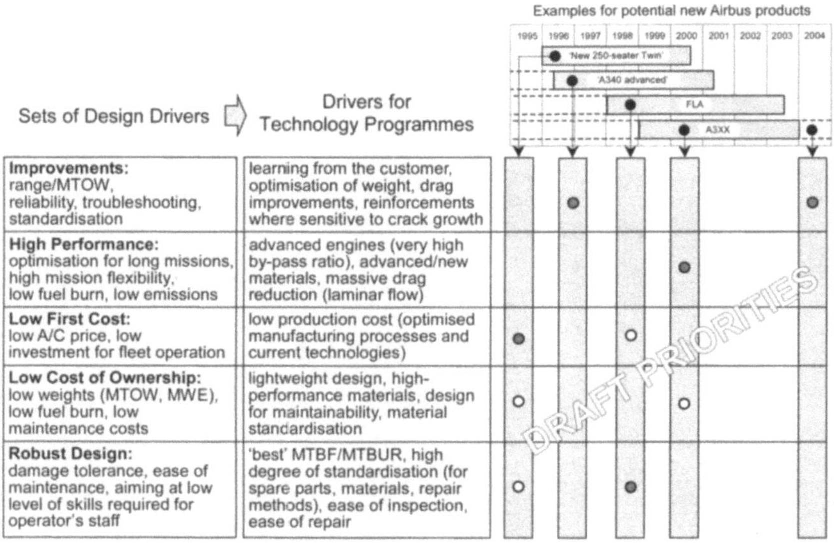

Figure 7: Potential future projects and technology drivers (established 1994).

The key question is now: 'How can the first priority be achieved without compromising the second priority and/or the remaining categories?' For the purpose of discussion, it is helpful to establish the so-called spider plot, see Figure 8, describing an initial global profile for each new product. First priorities are arranged next to the outer circle, neutral positioning is located on the intermediate circle and shortfalls are indicated where the mark is close to the centre. The initial judgement for the A3XX displays that *High Performance* can be achieved with support through the application of advanced and new technologies. A shortfall for *Low Cost of Ownership* is indicated in order to 'signal' that such technologies tend to increase operational costs through higher maintenance effort. A negative impact on the *First Cost* (i.e. airframe production costs, etc.) and on *Robust Design* also expresses major concerns related to the application of advanced/new technologies.

It is absolutely mandatory to recover from shortfalls indicated by the profiles. Thus, the technology down-selection process requires elements, which assure that the above concerns are addressed. The above profiles were designed to give global orientation for technology programmes.

Element: 'Profiles' for Material Candidates

Much more specific than the global profiles for future aircraft programmes are the profiles for new materials. The profile shown in Figure 9 delivers an initial judgement of benefits and penalties expected from Glare application in fuselage panels of the A3XX. Various parameters are clustered under the following categories: *First Cost*, *Maintenance Cost*, *Mission Costs* and *Mission Flexibility*. Glare is characterised by superior damage tolerance and associated weight savings as well as high corrosion resistance, see [1] for more details. Taking the operator's view for the material profile, Glare improves maintenance aspects and increases the resale value through high damage tolerance and corrosion resistance. Saving weight has a positive effect on mission costs and mission flexibility, resulting from less fuel burn and less emissions. A relatively small negative impact on aircraft price is also indicated by the profile, which was established in mid-1994. It must be admitted that the profile was extremely optimistic regarding the impact on airframe costs at that time, and it took six years of Glare development to get costs of this material candidate down to the expected level. In general, Figure 9 is still valid and thanks to co-operation with maintenance experts from key airlines, repair solutions have been designed into the A3XX structural concepts.

Element: Initial Set of Structural Design Drivers

In order to assure 'the best match of material characteristics with structural design drivers', the down-selection process requires that the main drivers for structural design be identified at a very early stage. An example is given by Figure 10, which delivers the main criteria for sizing of structural parts. Stronger requirements for corrosion resistance in fuselage bilge areas are also addressed. Such maps of 'structural design drivers' help to select material candidates.

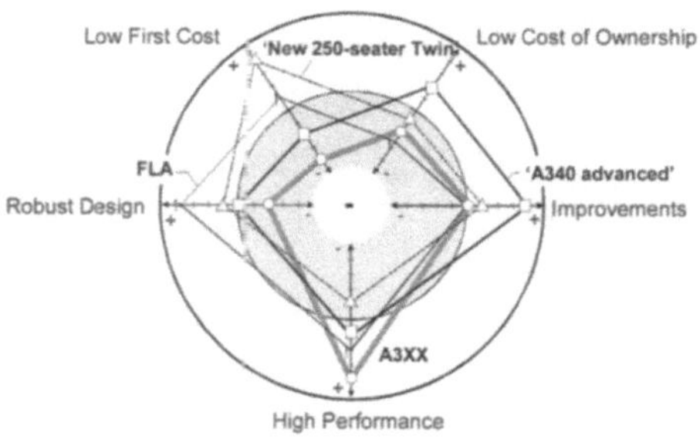

Figure 8: Global profiles for future aircraft programmes (year 1994).

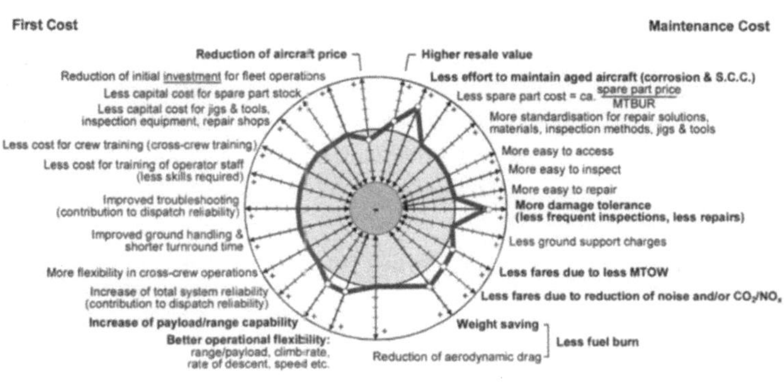

Figure 9: Profile for Glare, taking the operator's view (established 1994).

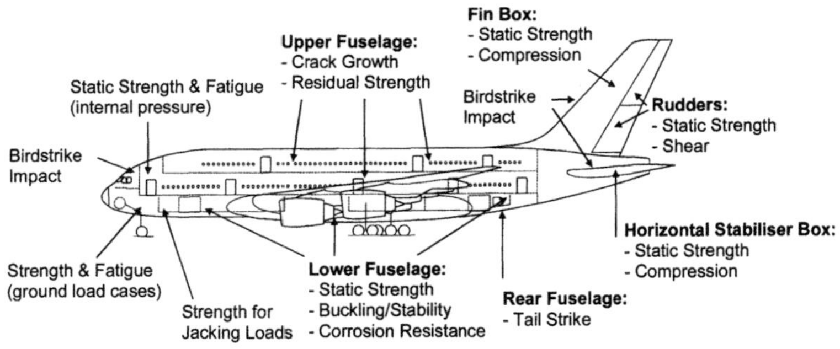

Figure 10: Structural design drivers for fuselage and empennage of A3XX.

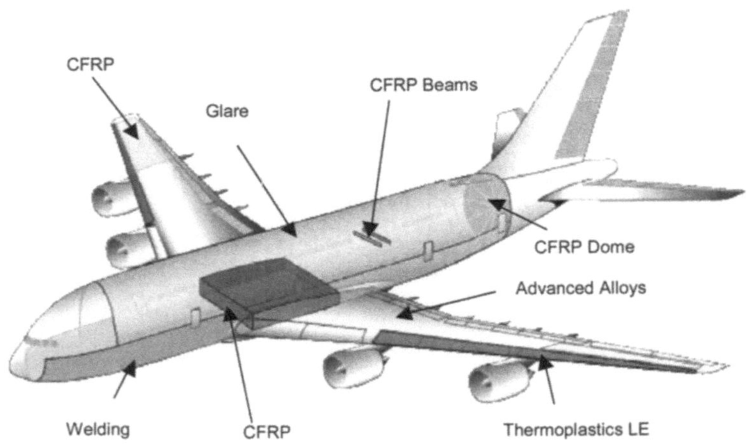

Figure 11: Initial scenario for potential material candidates.

Element: Reference and Initial Scenarios for Material Candidates

At the starting point of the evaluation of advanced/new materials, a so-called 'reference configuration' is established: state-of-the-art materials in combination with appropriate design concepts are chosen, supporting weight and cost estimates. At the next step, different scenarios with advanced/new materials and manufacturing processes are created. Then, all benefits or other implications can be measured against the 'reference configuration'.

The scenarios aim at first at 'the best match of material characteristics with structural design drivers', as discussed in the previous paragraph. Also, new manufacturing processes and their impact on structural design solutions are discussed on this basis, resulting in design alternatives, e.g. for structural joints and panel arrangements. One of the scenarios for the A3XX was established in 1997, see Figure 11. At this early stage of the programme, knowledge about loads/load path and stress distribution/levels was poor because of a lack of finite element analysis. Thus, the distribution of Glare application and welded panels, for example, was very much simplified compared to the definition status at the programme launch. The down-selection process is in essence a learning process, which delivers stepwise refinements of the 'initial scenarios for material candidates'.

Element: Analysis of Materials and Manufacturing Process Costs

In general, suppliers offer more advanced or new materials at a higher cost than existing materials. Years after initial introduction, prices tend to go down because production volume increases, allowing the supplier to recover initial investments. However, commercial aircraft manufacturers have to target cost reductions, even when more expensive materials are introduced. This target can only be achieved if advanced/new materials allow for new and less costly manufacturing processes. It has been outlined in Hinrichsen [2] that changes in manufacturing technologies can successfully support such an approach. Again, a reference is established, describing conventional technology. Taking fuselage panels in the upper centre fuselage as an example, Figure 12 compares Glare panels with the reference technology. Conventional panel production applies roll-forming and/or stretch-forming for stringers and skins. Glare panel manufacture is basically a layup of aluminium foils and pre-impregnated fibres in a mould, which shapes the final outer contour of such panels. Variations in the number of layers deliver local reinforcements. Both lower material waste and cost reduction through a self-forming process help to offset the higher material costs. Δ(NRC) covers the higher nonrecurring cost for Glare, mainly as a result of moulds, autoclaves and robots for automated layup. The results support *target cost* for Glare at the same level as that for conventional panels.

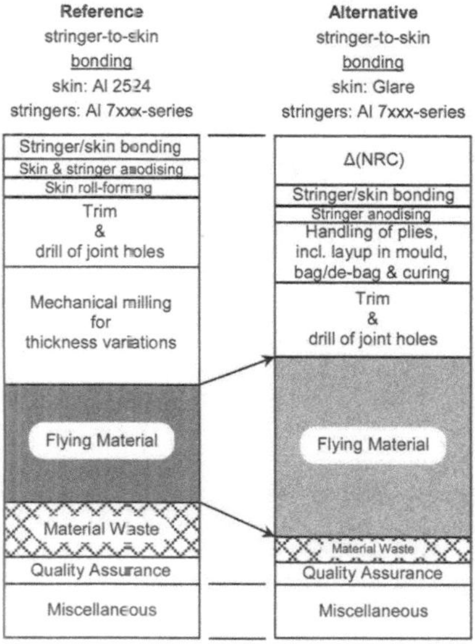

Figure 12: Fuselage panel cost comparison.

Element: Take Feedback from Airlines

All concepts for systems technologies as well as for structures and material application have been reviewed with experts from key airlines. For structures and materials, dedicated workshops have been held, bringing together maintenance experts from airlines with product support representatives and engineers working in the fields of material development and structural design at Airbus Industrie.

The main concerns addressed were: damage tolerance, inspection procedures and methods, accessibility, limits for size of repairs under airline responsibility, repairs using standard tools and standard materials, corrosion protection and prevention of galvanic corrosion whenever aluminium and carbon fibre materials have to be joined. Also, requirements for standardisation of materials, fasteners and repair solutions across the airframe were recorded as a major outcome of the common reviews. Full-scale repairs for representative structural parts helped to achieve airline acceptance. Figure 13 shows a flush repair for a damaged skin and stringer in a Glare panel as an example. Throughout the material down-selection process for the A3XX, no decision for the application of new materials was taken without a 'green light' from the airline experts.

Drilling for stringer coupling

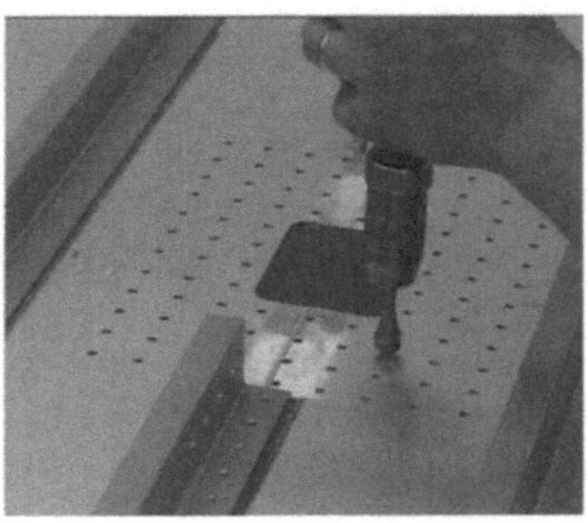

Deburring

Sealant application

Riveting

Figure 13: Standard repair for a Glare panel.

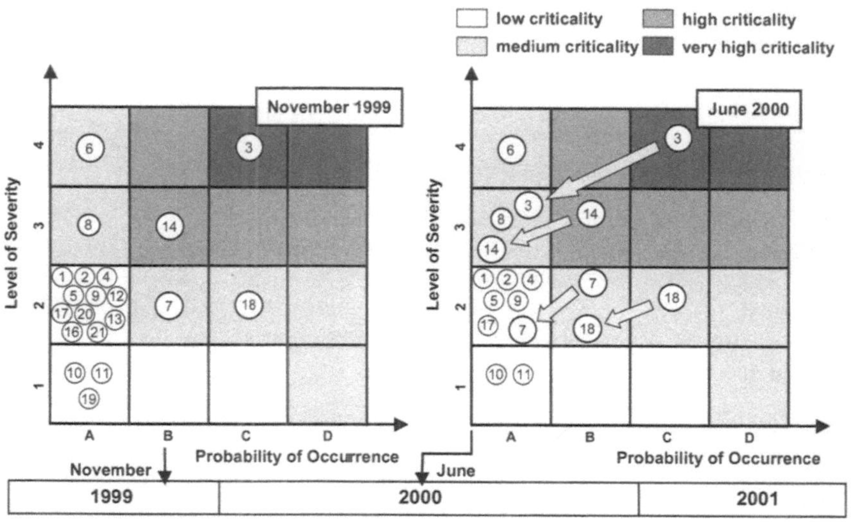

Figure 14: Tracking of risk mitigation using risk matrices.

Element: Risk Mitigation Plans and Risk Matrices

The initial application of new materials requires risk mitigation for the 'readiness' of those materials, their manufacturing processes and business conditions. Airlines have experienced problems whenever a new technology was put into service. In order to minimise the in-service problems and rework on production lines, risk mitigation plans were put in place. This process first establishes an initial list of potential risks. A judgement of the associated risks is classified by two categories: the 'probability of occurrence' and the 'level of severity'. The judgement for each item is than summarised in a risk matrix. Tasks are derived from the risk assessment and summarised in a risk mitigation plan, describing deliverables and target dates. The status of achievements is reviewed at certain milestones by the aircraft programme management, also incorporating senior experts, who are not involved in the specific tasks. Figure 14 gives an impression how risk items are driven down into the 'white' boxes of the risk matrix as time progresses. Unfortunately, the process is not always that stable, i.e. new risk items popped up while conducting the risk mitigation plan, either due to new input from airlines on maintenance issues or due to drawbacks in manufacturing trials. In reaction, recovery plans were launched.

Element: Initial Sizing

As shown in the schedule for material down-selection, see Figure 6, the initial sizing of aircraft components takes place during the last year of the concept phase. Stress calculations, stability and damage tolerance analysis are supported by finite element method models (FEM models) for major aircraft components and sets of loads data for the dimensioning load cases, driven by manoeuvres, gust and/or systems failure cases. Load calculations themselves are based at this stage of the programme on a complete FEM model for the primary structure of the airframe, see Figure 15.

Aerodynamic data resulting from numerical codes are adjusted to preliminary wind tunnel test results. Based on FEM model results, initial sizing can be performed for different material application scenarios, preparing the optimisation of the final material application.

Element: Optimisation and Final Freeze of Material Applications

Compared to the 'initial set of structural design drivers' described in a previous paragraph, progress is characterised through better knowledge about loads and load paths as well as stress distribution and elastic deformations. As an example, detailed FEM model analysis for fuselage panels and frames delivered information about high shear stresses above the wing/fuselage intersection and in the vicinity of the body/gear attachment, as marked in Figure 16.

As the initial scenario for potential material candidates in Figure 11 indicates, Glare panels were foreseen in these areas with very high shear stresses. Then the application in these areas had to be questioned because material development for Glare aimed at low crack growth rate and high residual strength. Today's material performance requires further development before a weight saving can be achieved in areas where shear loads drive the structural design. Consequently, alternative scenarios had to be investigated for the A3XX launch version.

Figure 15: Finite element model for loads and aeroelasticity.

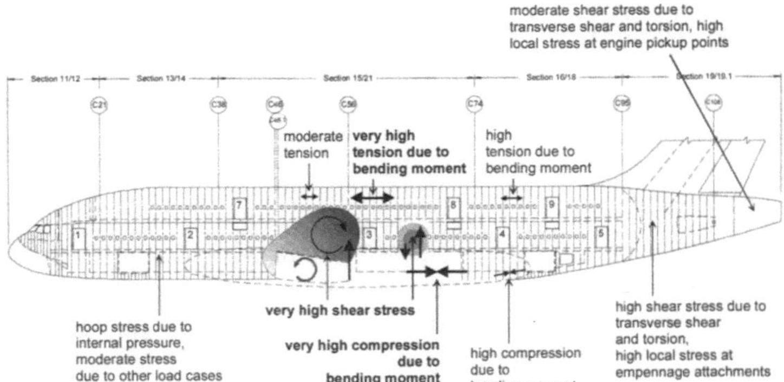

Figure 16: Results from FEM model calculations for fuselage panels.

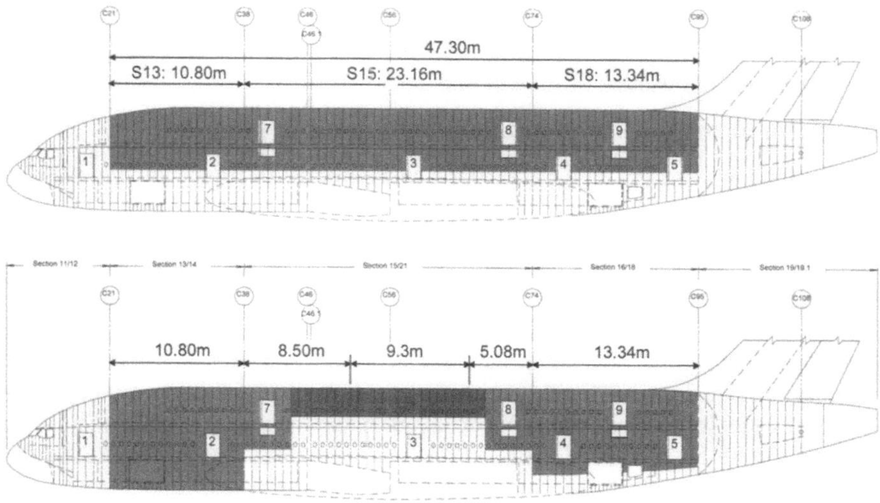

Figure 17: Alternative scenario for Glare fuselage panels.

In Figure 17 the initial scenario is plotted above an alternative scenario that was used to discuss the application of advanced aluminium alloys in those areas where Glare application does not pay off. Other alternative scenarios were also discussed in order to find the best solution that takes into account all relevant industrialisation aspects. Thus, the final optimisation is based on more than structural mechanics.

Element: Optimisation and Final Freeze of Manufacturing Processes

Another optimisation step at the end of the concept phase deals with the details of manufacturing alternatives. The down-selection process has prepared material application and manufacturing processes simultaneously and demonstrator programmes have been carried out. Requirements from structural mechanics have been settled and alternative material candidates have been evaluated, preparing the final selection of manufacturing processes. As an example, studies related to CFRP applications for panels of the aft fuselage and the tail cone are outlined below. In general, two automated layup processes are available as an alternative to hand-layup; Automated Fibre Placement (AFP) is as mature as Automated Tape Laying (ATL). Instead of placing a pre-impregnated unidirectional tape (80 to 300 mm wide), AFP machines work with up to 32 tows, placing them in a 150 mm wide strip in one shot. Each of these tows consists of a number of pre-impregnated fibres. The numerically controlled head of the AFP machine has a placement & cutting device for each of the tows, which enables the machine to achieve minimum gaps/overlaps in the layup and to follow very complex contours. These features deliver a proper *fibre placement* even on strongly double-curved surfaces of moulds, where ATL would fail because of unacceptable gaps/overlaps or because the tape layer would fail to give sufficient pressure across all of the tape to be placed. Compared to fabrics or tape, AFP allows the optimisation of fibre orientations to a larger extent.

To summarise the general abilities of AFP:

- automated layup on very complex surfaces
- maximum optimisation of fibre orientations
- automated layup on honeycomb cores

A restriction might arise with the size of the component to be produced in one shot.

The comparison of alternative manufacturing processes for the envisaged application for fuselage panels takes into account the following parameters: potential fibre orientation optimisation and complexity of geometry, see Figure 18. Multiaxial stressing of these panels requires little optimisation of fibre orientation. Thus fabrics, tapes and tows fulfil the structural design requirements. The complexity of geometry is related to the double curvature of the fuselage outer contour. Without additional panel joints, Automated Tape Laying fails to properly place the tapes. As shown in Figure 18, both hand-layup and AFP are feasible. The results of the study are summarised in Table 1. The final selection has to incorporate cost comparisons and quality aspects.

Table 1: Impact of parameters on selection of material and manufacturing process.

Parameters	Comments
Structural design	Moderate multiaxial state of stress Reinforcements at local load introductions
Material selection	Dry or pre-impregnated fabrics, tapes or tows
Degree of double curvature	Automated Tape Laying is ruled out because of penalties from requested additional panel joints
Component size	Close to limits with 4 panels in the aft fuselage and with 2 panels in the tail cone

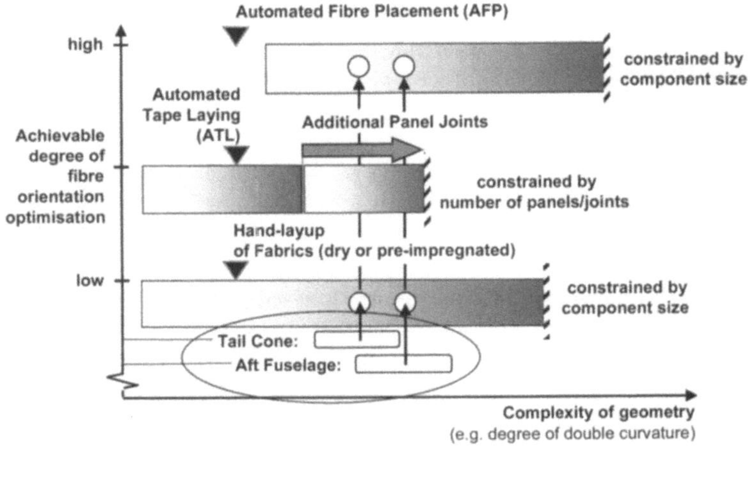

Figure 18: Comparison of alternative manufacturing processes.

Summary

Evaluation of advanced or new technologies urges us to look at the aircraft as an entity. All attempts to deal with a potential material candidate in isolation have failed so far. Scenarios, differing in materials and/or areas of application, can help to find the 'best compromise'. Also, 'the lessons learned' from the existing fleet must be transferred into new concepts and detailed design. Decisions for applications of advanced materials shall be in harmony with an evolution within the Airbus family. To step into a primary structure with a new technology first time requires special attention and careful preparation, including hardware tests and full-scale manufacturing demonstrators. Furthermore, such initial applications will take place under certain limitations, such as restricted ramp up in production volume, exclusion of areas with the highest load level, use of conservative margins for allowable stresses, and so forth. For a later version of the launch variant, such margins will be relaxed and areas of application increased. Margins are part of the so-called built-in potentials of a new technology.

The scenarios aim at first at the 'best match of material characteristics and structural design drivers', such as stability, damage tolerance, strength and stiffness. Also, new manufacturing processes and their impact on structural design solutions are discussed on this basis, resulting in new design alternatives, e.g. for structural joints, panel arrangements, etc.

Each step of the down-selection process delivers the input for an update of the aircraft configuration, which evolves from status to status over time. A decision to introduce a new material into the next aircraft configuration status goes along with a risk mitigation plan. This sub-process reviews the material readiness and identifies all development tests to be performed for verification of design solutions and manufacturing processes with regard to the selected area of application (structural component of the aircraft).

Technology down-selection (including verification) must be regarded as a time-wise process. As the aircraft configuration evolves, more and more analytic data for loads and calculation of stress distributions are made available, and in parallel, testing delivers more details. Based on the continuously improved knowledge base, an optimisation step for material applications has to be performed before the final aircraft configuration can be frozen.

Finally, manufacturing processes have to be chosen. To illustrate the driving parameters, alternative materials in combination with different manufacturing processes have been discussed for the non-pressurised aft fuselage and the tail cone.

A brief description of the complex down-selection process required some structuring and for this purpose, so-called principles and elements have been introduced. Clustering tasks, deliverables and schedules this way is not mandatory. The presented paper concentrated on technical issues and tools for the management of tasks. The paper is not complete, mainly because human factors have not been addressed, as 'technology is people', i.e. without their ideas, commitment and endeavour the challenges would not have been mastered.

Acknowledgement

First I would like to thank Jürgen Thomas for his continuous support during the years of technology preparation and concept development for A3XX. Jürgen challenged and stimulated us, the A3XX team, until success was achieved, i.e. programme launch for A380.

Thanks to all my colleagues in the Airbus community and to those working in research institutes and laboratories at material suppliers. Thanks to the experts from airlines, giving guidance to 'design for maintenance'.

Thanks to all who make it happen!

References

[1] J. Hinrichsen, 'Airbus A3XX: Design Features and Structural Technology Review', International School of Mathematics 'G. Stampaccia', 28th Workshop: *Advanced Design Problems in Aerospace Engineering*, Galileo Galilei Celebrations, Erice-Sicily, Italy, July 11-18, 1999.

[2] J. Hinrichsen, *Airbus A3XX: Materials and Technology Requirements*, 18th European Conference on Materials for Aerospace Applications, Association Aéronautique et Astronomique de France (AAAF), Le Bourget, France, June 16-18, 1999.

Response 1

Airbus composite aircraft fuselages - next or never

Michel J.L. van Tooren

Faculty of Aerospace Engineering
Delft University of Technology

Introduction

The application of composites in the pressurised fuselage shell structure has still not been accomplished for large civil transport (LCT) aircraft. The LCT aircraft industry seems therefore to be running behind the military and general aviation (GA) industry, where the application of composites in fuselage structures is becoming the standard.

The Airbus A380 will be the first aircraft in which partially non-metal skin materials will be applied. Large parts of the A380 fuselage crown panels will be made of Glare, a second-generation fibre metal laminate. It has also been decided not to use fibre-reinforced polymers as the basic skin material for the A380.

In this article, the background of the difference between material application in the GA and LCT aircraft industries will be analysed. The analysis results will be used to formulate conclusions and recommendations regarding the necessary development of material and production technology to achieve composite fuselages for the next generation of LCT aircraft.

Status of composite fuselage design

The GA industry has adopted composite technology to a considerable extent. Currently composite fuselages are common practice for non-pressurised 2-4 seat aircraft; they are available for pressurised 6-8 seat aircraft and planned for 10+ seat aircraft.

2-4 seat aircraft category

This category includes aircraft such as the Cirrusdesign SR-20, see Figure 1, and the Euro-ENAER Eaglet, see Figure 2. The fuselages of aircraft in this category are of the sandwich construction type. The facing material applied is glass-fibre-reinforced epoxy. The core material is foam, which can be PVC foam or PMI foam. The manufacturing methods vary from wet layup to prepreg technology. The fuselage sandwich skins are supported by a limited number of internal frames. Wooden and aluminium inserts are used for local concentrated load introduction. Typical basic layups are two-layer facings consisting of one 0/90 layer and one +/-45 layer.

6-8 seat aircraft category

This category includes aircraft such as the Extra 400, see Figure 3, the Beech Starship and the Raytheon Premier I, see Figure 4. The fuselages of these aircraft are of the sandwich construction type. The facings are made of carbon-fibre-reinforced epoxy. The core material is Nomex honeycomb. The manufacturing techniques applied, range from wet layup (Extra 400) to fibre placement techniques (Premier I). The fuselage skins of the Extra 400 are draped over an external skeleton made of frames and longerons. The skins of the Starship and the Premier I are supported with frames that are placed inside the fuselage sandwich panels. Due to the circular shape of the fuselage of these aircraft, there is no need for frames.

Figure 1: Cirrusdesign SR-20.

Figure 2: Euro-ENAER Eaglet.

Figure 3: Extra 400.

Figure 4: Raytheon Premier I.

10+ seat aircraft category

Several aircraft are planned in this category. In the GA section we find the Raytheon Hawker 450, see Figure 5. The fuselage of the Hawker 450 will be similar to the Premier I fuselage. This means it will have a cylindrical fuselage with sandwich skins and some frames inside the skin sandwich panels. These internal frames serve for the introduction of local loads from for example the floor. Within Airbus, a study into a full composite fuselage for the A320 sequel is being undertaken, see Figure 6. No decision has yet been made for the Airbus fuselage. Several concepts are under study. The monocoque fuselage as applied in the Hawker 450 is one of them.

Figure 5: Raytheon Hawker 450.

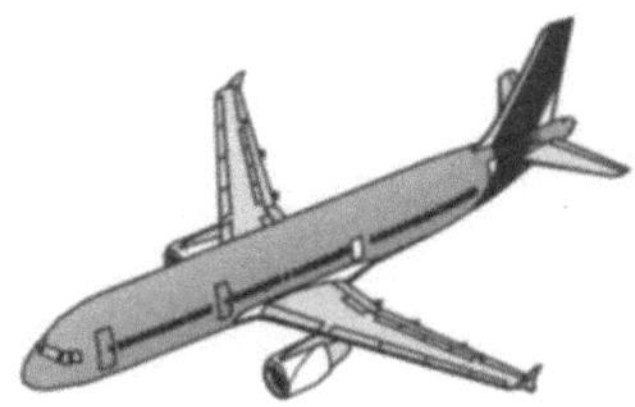

Figure 6: Airbus A3??.

Analysis of the current stage of composite fuselages[1]

In this section we study the reasons why composites are applied in GA aircraft. To do so, we first give an overview of the requirements that are applicable to fuselages. A schematic overview of all the requirements that have to be fulfilled is shown in Figure 7. In this diagram the requirements are divided into 1) structures and materials design and engineering requirements and 2) integrated manufacturing requirements. Improved integration is shown, travelling from the origin along the axes. Combining items on both axes yields design areas. The design areas can be used to classify the current fuselage designs. The current design practice for metal fuselage structures of transport aircraft is covered by the depicted *metals* and *metals + polymers* areas. The structural design covers not more than the minimum required, i.e. strength, stiffness, damage tolerance and producibility. All other requirements are fulfilled by non-integrated items, like isolation blankets.

In the composite fuselages applied for the 2-4 seat, the 6-8 seat and the 10+ seat category, all requirements are met in a single integrated design. The requirements listed on the horizontal axis, from basic strength to thermal and acoustical isolation, are met without adding elements to the basic fuselage shell. The same is true for the requirements listed on the vertical axis. The basic fuselage shell design fulfils the requirements ranging from producibility to integration of acoustical and thermal isolation.

[1] See [1-7].

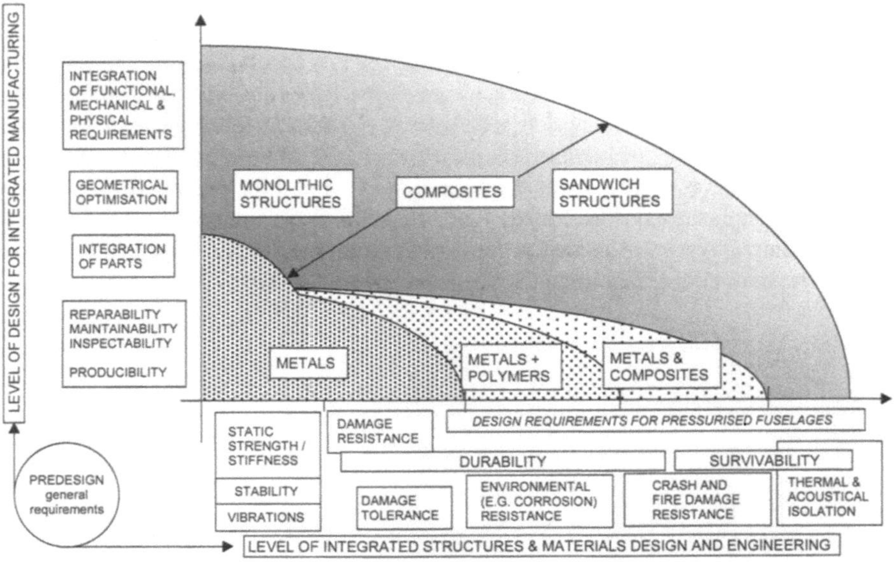

Figure 7: Design requirements for fuselage structures [2, 7].

The general requirements as indicated in Figure 7, are detailed in the JAR and FAR codes of the European and American Aviation Authorities respectively. They are quite similar for GA aircraft and LCT aircraft. Some of the aircraft examples mentioned earlier, like the Raytheon Premier I, are certified according to FAR 25, i.e. the same code as used for LCT aircraft. The difference in extent of composite application between the GA and LCT category can therefore not be explained by differences in the requirements.

The main improvements found and/or expected so far when applying composites in the fuselage are related to:

- Durability: Corrosion is no longer an issue when composites are applied as the main structural material. This not only helps to lower the maintenance cost, but also improves safety. Corrosion and fatigue damage occurring in relation to corrosion are the main issues of concern in metal aircraft.
- Survivability: A proper choice of resin system can result in fire resistance that will supersede the fire resistance of metal structures. Particularly for large aircraft this could mean that the 90-seconds rule for evacuation could be relaxed. Composites have excellent crash energy absorption characteristics. This is shown by the experience in Formula One car racing, but also by the limited experience so far with crashes of composite aircraft. Well-designed composite structures can absorb a large amount of energy without interfering too much with the volume designated to the user of the structure. This means that fuselage structures can be designed with a crash-absorbing lower part and

an occupant-protecting upper part. The lack of plastic deformation of composite materials is an advantage and not a disadvantage in this respect.

- Damage resistance and tolerance: Composites show excellent resistance against cyclic loading. Fatigue is therefore no longer a design driver in the fuselage design when composites are applied. In general the design strains are low enough to ensure that in the case of manufacturing or in-service damage (impact), the no-growth principle can be applied to the primary structural elements (PSEs) of the fuselage, see Figure 8. In the latest certification programmes, however, the slow damage growth philosophy has also been applied for inspectable PSEs.
- Reparability: The composite structure can be easily repaired. Not only will strength be regained but also other properties like the aerodynamic shape. Therefore it is better to talk about 'restoration' than repair. This is especially true for the GA aircraft. The monocoque fuselages have thin facings that can be restored with simple techniques. Bonding a patch with a certain overlap over the edges of the area containing the damage normally restores the damaged area.
- Parts integration: The manufacturing methods applied for composite structures allow for the integration of parts to a very high level. This means that no fasteners are required, which reduces both material and labour costs. In addition, the absence of fasteners allows for noiseless assembly. Riveting during assembly of metal sheet structures is often a very noisy affair.
- Geometrical optimisation: The fact that composites have a very good drape during part manufacturing allows for complex shapes. Double curvatures and other complex geometries can be made without the need for excessive forming forces or pressures needed in metal sheet forming processes.
- Integration of physical and mechanical properties: The application of composites as a monocoque structure in the aircraft fuselage allows for the integration of physical and mechanical properties. This is an added benefit up to now but could be exploited more extensively in transport aircraft.

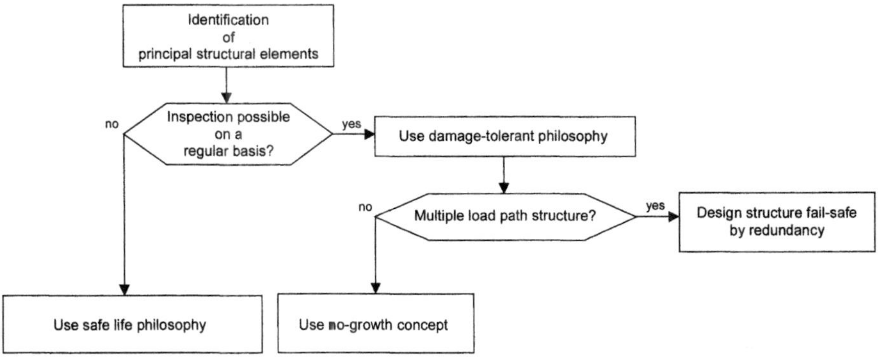

Figure 8: Design philosophy decision chart for composites.

Considering all the advantages of composites mentioned above, it is remarkable that even the latest generations of civil transport aircraft do not have a composite fuselage structure. It is not that the advantages of composites are denied by the aircraft industry. This can be seen in Figure 9, where the growth of the application of composites in Airbus aircraft is shown. The rear bulkhead and the keel beams are the first principal structural elements in the fuselage that are made of composites in the latest generation of Airbus aircraft. However, composites have still not been applied in the fuselage shell structure. The reasons why it seems important to change this situation have been discussed above and are summarised in Figure 10. From this figure we can conclude that the LCT aircraft industry could seriously benefit from a wider application of composites.

In order to get a clear picture of the differences between the aircraft currently flying with composite fuselages and LCT aircraft, the three elements that define a structural concept, i.e. material, shape and manufacturing technique, are compared for both categories in Figure 11. From this overview we can conclude that the LCT aircraft industry applies a wider range of materials and manufacturing techniques than the GA industry, and that the materials applied in the LCT aircraft industry include the materials and processes applied for GA aircraft. So it still remains unclear what the differences result from. Most likely the difference must come from the absolute size of the structure and the risks involved in designing and operating the structures.

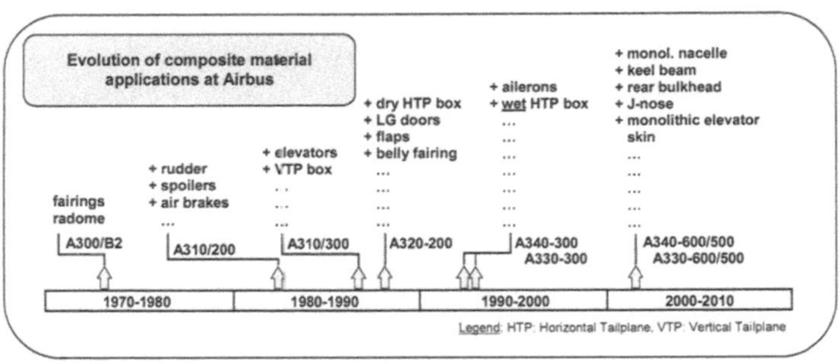

Figure 9: Application of composites in Airbus aircraft.

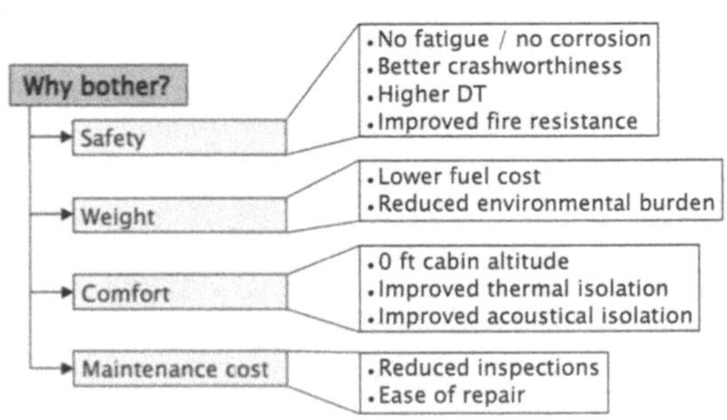

Figure 10: Why bother about applying composites in aircraft fuselages?

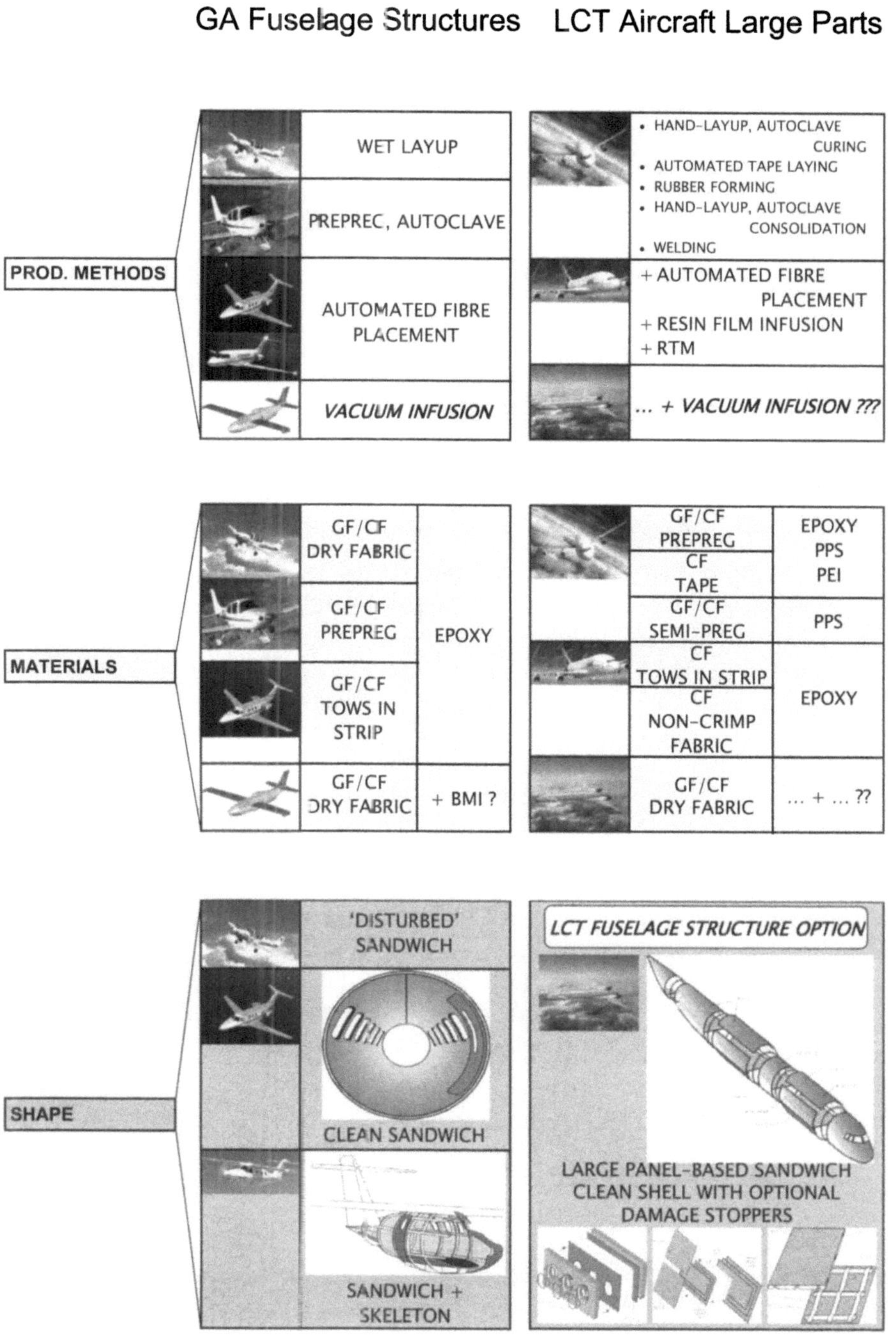

Figure 11: Comparison between GA and LCT aircraft structure design and manufacture practice.

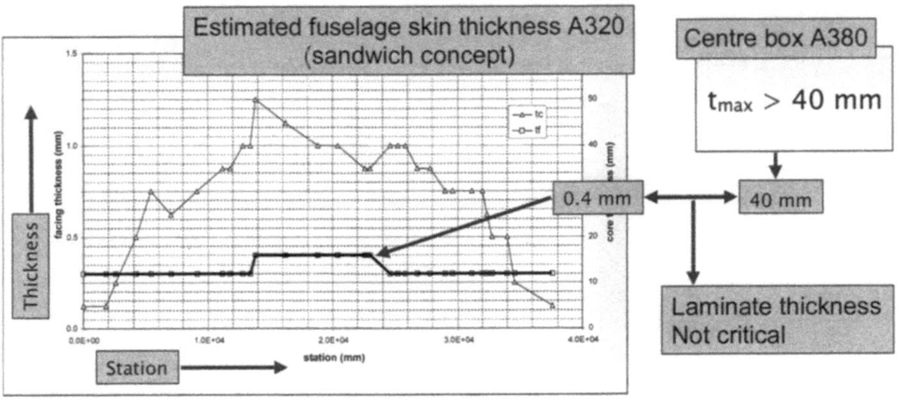

Figure 12: Thickness of laminates in monocoque sandwich LCT aircraft fuselage compared with A380 centre box design [6].

Let us start with an analysis of the size. The fuselage shells in the GA aircraft have limited dimensions. The facings of the applied sandwich shells have a thickness below 1 mm. The structure will behave very much like a plane-stress design, which is important to prevent composite failure modes originating from through-the-thickness stresses. In Van Tooren [6] a rough estimate is given for the thickness of the facings of a fuselage for an A320 type aircraft. It appears that the required thickness for stability and strength is also below 1 mm. The LCT aircraft industry applies and will continue to apply structures, like the centre box of the A380, with laminate thickness up to 40 mm. Therefore, the fuselage shell thickness does not explain the absence of composite fuselages in the LCT category, see Figure 12.

The other dimensions of importance are the length and diameter of the fuselage. With respect to the design one can generally say that larger structures do not cause bigger problems than smaller structures. In general the LCT aircraft have relatively more 'undisturbed' structure than GA aircraft, which makes the design easier. This is different for the manufacturing aspects. The size of LCT aircraft will create problems related to:

- accessibility of tools during manufacture
- the 'pot life' of the applied polymers
- the required size of the curing or consolidation equipment

GA aircraft have a size for which accessibility in production can be handled through clever structural division and mould design. For LCT aircraft much more far-reaching divisions have to be made.

Fuselages for LCT aircraft

A possible solution for a monocoque fuselage of a LCT aircraft is shown in Figure 13. A limited number of longitudinal and circumferential divisions yield shells that can be made in readily accessible moulds. Clever design of the joints could yield production without the use of autoclaves.

For the individual shells and the assembly of these shells, several manufacturing technologies can be applied. Most currently applied techniques that are used for the production of composite parts for LCT aircraft, are prepreg-based and require an autoclave. This (partly) open mould manufacturing technique has a number of drawbacks; high cost, both recurring and non-recurring, are an obstacle for further application.

Several alternatives are available. An increasingly important family of manufacturing techniques is called resin transfer moulding (RTM). So far the aircraft industry has mainly focused on RTM with stiff (steel) moulds, in which resin is infused into the moulds with high pressure (± 5 bar). In other market sectors (ships and wind turbine blades) a vacuum infusion technique has been developed for the production of large parts (± 0.3 bar). Airbus uses the film infusion technique – called 'film stacking' in the past – for the production of the rear bulkhead of the A340-500/600.

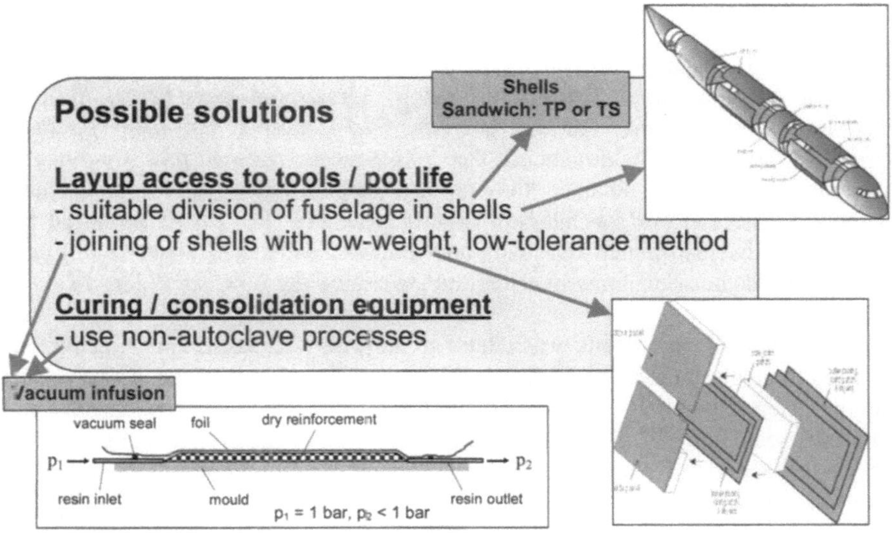

Figure 13: Possible concept for LCT aircraft composite fuselage.

Figure 14: a) Preparation for injection, b) first part of injection and c) halfway injection.

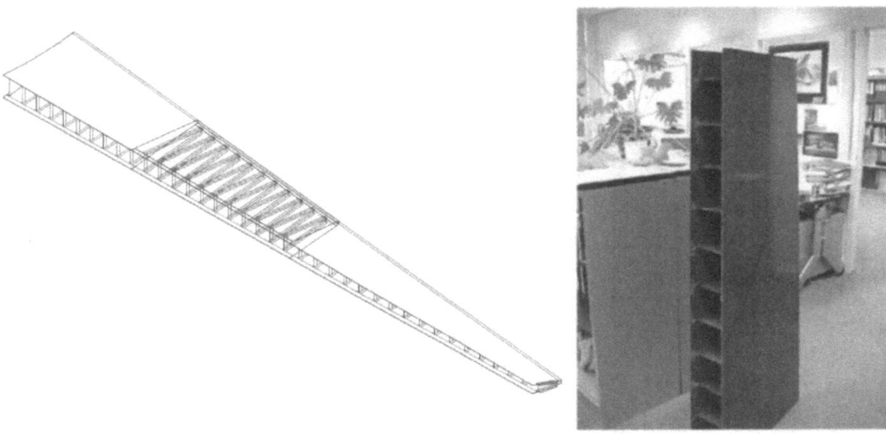

Figure 15: Multi-rib design with integrated top and bottom skin and ribs.

Recently, quite successful research was completed on the application of the vacuum injection – or vacuum infusion – technology for aircraft parts [8, 9]. Infusion experiments were carried out on two structural concepts, i.e. foam sandwich structures and multi-rib structures. The experiments showed that good-quality sandwich panels can be obtained. Flow through both facings appears simultaneously without the necessity of special flow control measures. The panels produced had different sandwich-stiffened areas with non-sandwich areas in between. In this case, there were also no special measures required to control the flow, see Figure 14.

The second set of experiments was related to multi-rib structures. The objective was to show that the integration of ribs, spars and skin is feasible and no secondary bonding step is required. As a demonstrator part, a piece of a control surface was selected, see Figure 15. The initial experiments have clearly pointed out some advantages and disadvantages of the application of vacuum infusion technology for large aircraft parts.

Benefits are:

- <u>High level of integration of parts is possible</u>: Compared to prepreg autoclave manufacturing it is easier to achieve a high level of part integration, which lowers the assembly cost considerably.

- No special storage requirements: Storage demands little attention. The fabric can be stored in a dry room and the resin and hardener each in a separate chemical storage facility. For prepregs a freezer is needed to prevent the prepregs from curing. Prepreg storage is therefore relatively expensive.
- No expensive equipment required: For infusion and curing a part with vacuum infusion only an oven is needed. For manufacturing a prepreg part an autoclave is necessary which is expensive compared to an oven.
- Use of simple moulds and tooling: The vacuum infusion production uses simple tools and moulds whereby process corrections are possible, within limits, during the infusion. Although the same product can be made with prepregs it will be difficult to achieve the same level of integration whilst keeping the tooling simple. The process cannot be corrected during curing.
- No time-critical positioning of prepreg/reinforcement: The reinforcement for the whole product can be placed in the mould without any time restraints. Time becomes critical when the resin and hardener have been mixed and the product has to be infused. For prepregs the available time (open pot life) is limited after the prepregs have been placed outside the freezer, since they will start to cure. In this time the prepregs have to be cut, placed in the moulds and prepared for the autoclave procedure.

Drawbacks are:

- Fibre volume content is more difficult to control: When using prepregs the fibre content is optimal and well controlled. Prepregs are pre-impregnated with the optimal amount of resin. In vacuum infusion with flexible tooling the fibre content can depend on the pressure (vacuum) applied and the location in the part (local pressure).
- Finishing can be more time-consuming: After infusion some post-treatment is required to remove resin-rich areas and to establish the required surface quality on specific locations, such as areas for secondary bonding. Due to the use of pre-impregnated prepregs and a well controlled pressure (also locally) the surface quality and thickness on the foil side is better controlled.

Research and development in the following fields is required to come to actual application of vacuum infusion technology for complex parts in LCT aircraft:

- multiaxial fabrics to reduce layup cost and facilitate production preparation
- preformed custom-made vacuum bags to ease sealing and improve production speed
- improved methods for preform fixation
- quality control on multi-material products

Jointing techniques based on vacuum infusion need further development to be applied for fastener-less assembly.

Other techniques suitable for certain elements in the fuselage structure are also becoming more mature. Thermoplastics like PEI and PPS, processed with pressing techniques, prove themselves suitable for parts where damage resistance is important or where the number of identical parts is substantial. One could imagine part of the fuselage shells being made as thermoplastic sandwich panels. Infusion technology or welding could be applied to join these individual shells. Film infusion has been shown to be a cost-effective technique. Also, this technique can be combined with other forms of injection technology for the assembly.

It is important that assembly technologies are developed. Mechanical jointing of composite parts is a rather cost- and weight-ineffective approach. Clever fastener-less assembly could develop competitive structures. Experiments have been performed with infusion jointing of composites. Very promising results have been obtained. More research is needed in the multi-step infusion of composite structures. The quality of co-injected and sequentially injected parts is still an unknown area.

The use of monocoque structures – both monolithic and sandwich –, the application of new manufacturing techniques and the required changes in assembly methods bring forward the need for new design rules and tools. Sandwiches have a bad name in the LCT aircraft industry. This is mainly due to maintenance problems with ill-designed control and high-lift surfaces – water ingression in products with Aramid skins and poorly processed skin/core bonds in products with carbon/epoxy skins. It will be the challenge for the designers to show that these problems can be overcome with innovative and proper design.

Conclusions

It can be concluded that the general aviation industry is ahead with the application of composites in pressurised fuselages. The reason for this cannot be found in the difference in requirements. Most likely the difference in size between GA aircraft and large civil transport aircraft structures is the main cause. The size of LCT aircraft yields problems with respect to manufacturing and costs.

Manufacture and assembly of large structures outside the autoclave becomes necessary for LCT aircraft fuselages. For part production, vacuum infusion and pressing techniques are candidate technologies. Designing for the infusion technology is an important condition for successful application. This not only implies geometry considerations and a level of integration, but also proper input data for detailed design calculations. Therefore, effort needs to be spent on determination of achievable material properties with the selected materials and the process during the design stage of the part.

With the change in design and production philosophy, full composite pressurised fuselages will be next.

References

[1] Th. de Jong, A. Beukers and L.B. Vogelesang, 'Weight Reduction as an Added Benefit', in: *Fatigue of Aircraft Materials*, Proceedings of the Specialists' Conference, Delft: Delft Structures and Materials Laboratory, 1992, ISBN 90-6275-809-6: 199-214.

[2] A. Beukers and M.J.L. van Tooren, 'Ontwerp-filosofie van de Extra 400 koolstof romp' (in Dutch), in: *De Constructeur*, 10, October 1997.

[3] M.J.L. van Tooren, 'A new step to easier production of high quality sandwich structures', in: Proceedings of the first conference on sandwich constructions, Stockholm, Sweden, June 19-21, 1989: 577-597.

[4] M.J.L. van Tooren, M.N. van Beijnen and I.P.M. van Stijn, 'Towards an all composite aircraft fuselage', in: Proceedings of ICCM/9, Vol. 6, Madrid, Spain, July 12-16, 1993.

[5] A. Beukers, 'Cost Effective Composite Plate and Shell Structures for Transports by Manufacturing Technologies like In-situ Foaming, Thermoforming and Pressforming Continuous Fiber Reinforced Thermoplastic Sheets', in: Proceedings of the 13th International SAMPE Conference, European Chapter, Hamburg, Germany, May 11-13, 1992.

[6] M.J.L. van Tooren, and A. Yoshii, 'Study on the merits of ACM sandwiches for aircraft and automotive structures and some recommendations for improvement', in: Proceedings of the 2nd Japan International SAMPE Symposium, Tokyo, Japan, December 11-14, 1991: 1152-1159.

[7] M.J.L. van Tooren, 'Sandwich fuselage design', Delft University Press, 1999.

[8] M.J.L. van Tooren, M.P. Dirven and A. Beukers, 'Vacuum Injection in Aviation Manufacturing Processes', in: *Journal of Composite Materials*, Volume 35, No. 17, 2001: 1587-1603.

[9] A. Hoebergen, A. Brødsjø, M.J.L. van Tooren and M. Verhoeven, *Development of a vacuum infusion process for aircraft part manufacturing*, TNO report 486-000818-1ahr, 2000.

The way to ensure technology maturity for new materials: A contribution to airworthiness issues

Jean Rouchon

Toulouse Aeronautical Test Centre CEAT
DGA

Joint Aviation Authorities

Abstract – This paper has been prepared for a conference organised in Delft on September 24-26, 2001 and entitled 'Glare - The New Material for Aircraft'. It is a contribution to the keynote paper presented by Jens Hinrichsen in a session entitled 'The material down-selection process for A3XX'. Its purpose is to supplement the paper by the aircraft manufacturer, by emphasising the airworthiness issues associated with this topic, with the views of a specialist working for a certifying agency. Airworthiness standards directly addressing material issues are reviewed and their interpretation is commented on in the light of the author's experience, working for sixteen years as JAA assistant specialist for the certification of composite structures in all the Airbus programmes from the A320 to the most recent versions.

Introduction

Airworthiness can be defined as the capability of an aircraft to be operated in proper and safe conditions. Rules, procedures and standards covering all aspects contributing to the safety of transportation by aircraft, began to be established in the 1920s and are continuously updated, taking 'lessons learned' into account, as well as the introduction of new technologies, operation constraints, etc. Minimum internationally recognised airworthiness standards have been defined by the International Civil Aviation Organisation (ICAO) since 1944. With these standards as a baseline, two main codes have been developed in the world, i.e. the Federal

Aviation Regulations (FAR) in the United States and the Joint Aviation Requirements (JAR) in Europe. They differ by some variants, but the best is done to keep these codes as close to each other as possible and for them to follow the same construction principles.

As far as structural integrity is concerned, three different issues have to be covered, i.e. design, production and maintenance. The relevant airworthiness standards for designing large aeroplanes (transport category aeroplanes) are provided by JAR – or FAR – part 25. Production is covered by JAR – or FAR – part 21, entitled 'Certification Procedures for Aircraft and related Products and Parts', while in the FAA code maintenance is covered by part 43, entitled 'Maintenance, Preventive Maintenance, Rebuilding and Alteration'.

Certifying an aeroplane is a process whereby a certificate of airworthiness is delivered by a state to a product when the applicant has demonstrated, and the state has verified, that this product complies with airworthiness requirements defined by the state. It is a fundamental assumption to state that complying with the requirements, plus adequate proficiency and professional integrity of all the actors involved in the design, the production and the operation of the aircraft is sufficient to guarantee an acceptable level of safety.

Materials issues in airworthiness standards

Since materials engineering properties govern most structural behaviours, it can be argued that materials aspects are directly or indirectly included in most of the paragraphs covering structure airworthiness requirements (static strength, fatigue/damage tolerance, flutter, birdstrike, continued airworthiness, etc.). However, there are only three paragraphs in the airworthiness standards where the word 'materials' actually appears in the title and which specifically address issues concerning materials. These are:

- § 25.603 - Materials
- § 25.613 - Material Strength Properties and Design Values
- § 21.125 - Production Inspection System - Materials Review Board

This paper will only comment on these three regulatory paragraphs, focusing on their interpretation with respect to the introduction of a new material in an aircraft application. Before these regulatory paragraphs are reviewed, the building block approach, as an adequate methodology to show compliance with the structure airworthiness requirements, will be detailed.

The building block approach

Illustrated by the pyramid of tests, see Figure 1, the building block approach provides the most appropriate view of the structural substantiation general principles, which are: 'analysis supported by test evidence.' From the measurement of intrinsic material properties at the bottom to the structure full-scale test culminating at the top, the pyramid of tests is an incremental process calling for an increasing specimen complexity.

Five main levels can be defined in this pyramid as a function of their purposes or finalities:

1. generation of material allowables at the bottom of the pyramid, where the intrinsic material properties needed to size the structure are processed on a statistical basis, in order to generate mean and either A- or B-values
2. development or validation of calculation rules for generic design features (e.g. a filled or an open hole), checking and/or generation of allowables at this design feature level of complexity
3. development or validation of calculation rules for non-generic design features, checking and/or generation of allowables at this design feature level of complexity
4. preliminary validation of the overall structure sizing for parts showing low accessibility to calculation; this may be done either to mitigate the risks of the programme by investigating in advance the performances of new materials/concepts or to assess the actual margins for those details that will not be brought to rupture in the final full-scale test

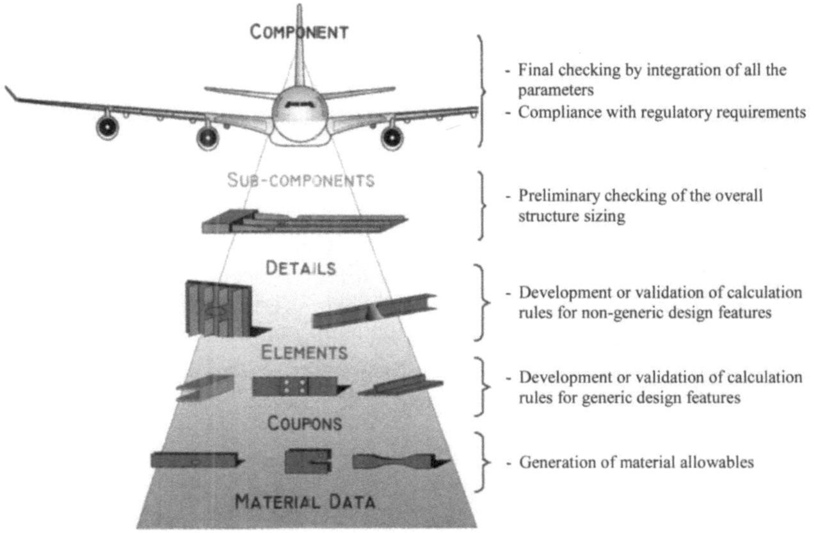

Figure 1: Pyramid of tests.

5. full-scale test for the final checking of the structure model and sizing, integrating all the parameters and showing compliance with the regulations when such a test is required (e.g. as per § 25.307 for static testing)

The advent of composite materials started the rise of the building block approach, and the expensive testing associated with the numerous and complex specimens illustrated by the pyramid of tests has often been questioned by programme managers responsible for the budgets. The reasons why more tests are required with composites are easy to understand and are listed below:

- Material anisotropy: More intrinsic material properties need to be measured at the coupon level.
- Material sensitivity to environmental conditions (mainly temperature and humidity): Matrix-controlled properties need to be measured in both as-received and aged conditions.
- Low accessibility to calculation for some complex design features: Failure modes may be complex and criteria poorly developed. Therefore, there is a need to generate or to check design values at a high level of specimen complexity.
- Material scatter: There is a need to increase the sample size in order to reduce the penalty inflicted by the statistical reduction of test data for the derivation of the allowables. In the simplest method to calculate an allowable (A- or B-value), the result is equal to the estimate of the mean, minus k times the estimate of the standard deviation. Since k decreases as the sample size increases, increasing the sample size and decreasing k accordingly can counterbalance the reduction inflicted by an elevated standard deviation value.

Although widely mentioned with composite materials, the building block approach applies to whatever the material and/or process and has implicitly been used in the past with conventional metals. As a result of such methodology, structure sizing or dimensioning is fully under control. This means that any change in the inputs (loads, material properties, etc.) has predictable effects. This is very useful in the situation of addressing a concession, certifying a derivative of the aircraft with increased loads, substantiating any change in the material or the design, etc. As a consequence, this building block approach should be carefully applied each time a new material/technology is introduced. Abundant testing is needed at all the levels of the pyramid at the early stage of the development. Then, as the understanding of structural behaviour and accessibility to calculation improve, the amount of testing needed to support the analysis will gradually decrease.

§ 25.603 - Materials

The content of this regulatory paragraph is:

The suitability and durability of materials used for parts, the failure of which could adversely affect safety, must:

(a) be established on the basis of experience or tests,

(b) meet approved specification that ensure their having the strength and other properties assumed in the design data, and

(c) take into account the environmental conditions, such as temperature and humidity expected in service.

The following comments will only focus on the sub-paragraph (b) of this regulatory paragraph. In its volume I 'Guidelines', MIL-HDBK-17 provides a useful definition of material qualification testing which is 'to prove the ability of a given material/process to meet the requirements of a material specification'. In other words, it is a regulatory requirement that all materials used in the production of an aircraft are qualified, which means that appropriate specifications must exist for this purpose. There is no problem with conventional materials/processes (light alloys, composites) for which each manufacturer has developed and owns its proper set of qualification specifications, but nothing adequate may exist, at the beginning, for a new and different generation of material. It is the opinion of the author that this process of establishing the original specification values should be addressed as early as possible, even though all new material potentialities have not yet been attained and there is only one supplier candidate to the qualification. In order to support these views, it is important to call up what the qualification process is intended to ensure:

- Engineering properties of the material/process, allowing for long-term behaviour, are sufficient with respect to the applications that are envisioned.
- Material presentation and physical properties comply with the manufacturer's projects and its workshop capabilities.
- The material does not exhibit any questionable features or properties, e.g. unfriendly chemical components with associated health hazards, unforeseeable behaviours, etc.
- Material fabrication key parameters have been identified and toleranced. A quality assurance system has been implemented that will ensure the consistency of material performances, which has been shown through the evaluation of several different batches.

As a result of this last point, the 'configuration' of the material and its associated performances are definitely frozen after the qualification process is achieved, and no uncontrolled deviations should then be expected. This is necessary to make sense of the work performed in the scope of the type certification, where it is fundamental that the materials and processes that will be used in serial production are representative of those used to manufacture the specimens for structural substantiation and which are illustrated by the pyramid of tests. Therefore, qualification work should be completed as soon as possible, ideally before structural substantiation testing, and the appropriate specifications being written down. This despite the fact, as already mentioned, that there may be only one material candidate to meet this specification. In addition to the fact that the 25.603 regulatory paragraph makes it mandatory, material qualification specifications are necessary to address second sourcing when opportunities will occur.

To close this paragraph, dedicated to the interpretation of JAR 25.603, it is important to point out some limits of the material qualification, which are listed below:

- While a material may be qualified to a given specification, it must still be approved for use in each specific application. In other words, qualification is a perquisite but not a sufficient condition to approve a material in view of any application.
- The generation of design values or allowables should not be the purpose of a qualification programme, except for those intrinsic materials properties and generic design features illustrated at the lowest level of the pyramid of tests.
- Qualifying a material is the manufacturer's own liability and can only bind him.

§ 25.613 - Material Strength Properties and Design Values

The content of this regulatory paragraph is:

(a) *Material strength properties must be based on enough tests of materials meeting approved specifications to establish design values on a statistical basis.*

(b) *Design values must be chosen to minimise the probability of structural failure due to material variability. Except as provided in sub-paragraphs (d) and (e) of this paragraph, compliance with this paragraph must be shown by selecting design values which assure material strength with the following probability:*

(1) *where applied loads are eventually distributed through a single member within an assembly, the failure of which could result in loss of the component, 99 percent probability with 95 percent confidence, and*

(2) *for redundant structures, those in which the failure of individual elements would result in applied loads safely distributed to other carrying members, 90 percent probability with 95 percent confidence.*

(c) *The strength, detail design and fabrication of the structure must minimise the probability of disastrous failure, particularly at points of stress concentration.*

(d) *Material specifications must be those contained in documents accepted by the authority.*

(e) *Other design values may be used if a selection of the material is made in which a specimen of each individual item is tested before use and it is determined that the actual strength properties of that particular item will equal or exceed those used in the design.*

This paragraph regulates the level of confidence of the design values used to size the structure, as a function of the structure design, for either single or multiple load paths. A useful definition of design values is provided by the Advisory Circular AC 20-107A, dedicated to composite materials, where the following is written:

Design values - material, structural element and structural detail properties that have been determined from test data and chosen to assure a high degree of confidence in the integrity of the completed structure,

while in the same documents, allowables are defined as follows:

Allowables - material values that are determined from test data at the laminate or lamina level on a probability basis, e.g. A- or B-values.

Material strength properties are random variables from which allowables can be statistically derived to represent, with a 95 percent confidence, an estimate of the 1st (A-value) and 10th percentile (B-value) respectively. Thus, for a given material property there are as many allowables as data sets, but only one design value selected for dimensioning. If several material sources are qualified and interchangeable for application in a component, one design value encompassing the allowables associated to each material source is defined. Figure 2 illustrates these definitions and includes the margin, which represents, for instance, the difference between the maximum stress at ultimate loads and the design value for static loading.

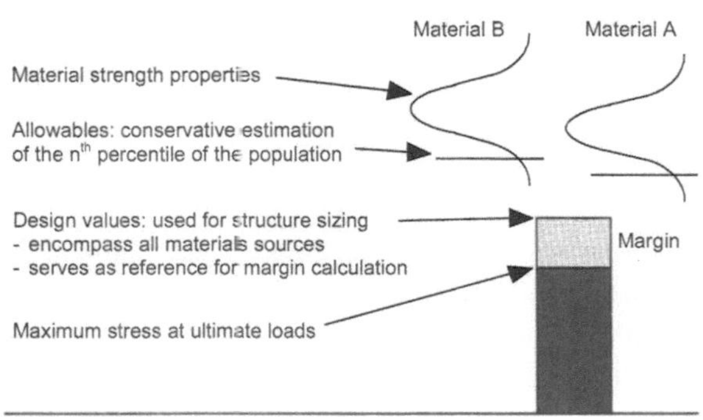

Figure 2: Definition of design values, allowables and margin.

No additional margin between design values and allowables is required by the regulations. However, the experience of the author has shown that choosing design values strictly equal to the allowables may be the source of potential issues for the future. Such tendencies generally come from the race to weight saving and the competition between new technologies and common – most often cheaper – ones.

One of these issues concerns the situation of a material's second-source qualification. Unless the second source can provide higher allowables, the certification documentation is to be reviewed to update margins, despite the fact they remain positive. Giving some room between design values and allowables facilitates the introduction of second-source materials or the way to address any deviation in the performance of the original material, which is another issue.

The second point to be raised in relation to this regulatory paragraph is the choice of the structural detail complexity to derive allowable values. Certification requires that margins are calculated at all critical points of the structure and are shown to be positive. The reliability of the demonstration will depend on:

- the right knowledge of the in-service loads and environmental conditions
- the quality of the structure's numerical or analytical model
- the quality of the failure criteria used to calculate the margin
- the reproducibility of material properties and processes

This last point is the combination of the purposes of § 25.603 and § 21.125 and has already been addressed above. Despite their real importance, the first two points are out of the scope of this paper. The following comments will therefore focus on the third point, which is the quality of the failure criteria in terms of representativeness and accuracy.

The general form of failure criteria that are applied to calculate the margins can be written down as follows:

$$F(p_1, p_2, \ldots, p_i; s_1, s_2, \ldots, s_i; t_1, t_2, \ldots, t_i) = 1$$

with: $p_1, p_2, \ldots, p_i$ = intrinsic material or structural detail properties
$s_1, s_2, \ldots, s_i$ = stresses, strains or loads
$t_1, t_2, \ldots, t_i$ = tuning factors

Tuning factors are determined by tests and are aimed at representing the influence of the design configuration on the failure criteria. The fatigue quality index FQI is a well-known example of a tuning parameter. Obviously, testing different and representative design configurations is needed to derive these tuning factors.

Ideally, the actual variability of the failure strength of the design point, where the margin is calculated, should be reflected by the allowable values. This can be achieved by actually testing several identical specimens fully representative of the design point or by using reliable failure criteria. The first choice is restricted to those structural details having no accessibility to calculation and is often very expensive since those details are in general of complex shape. The second choice relies on structure stressing and calculation with failure criteria where the number of tuning factors will depend on this accessibility to calculation. Understanding failure phenomenon and developing adequate failure criteria able to correctly reflect the actual value and variability of the point design performance is a very important issue governing airworthiness, but it may not be achieved at the early stage of the development of the new material/technology. Therefore, generating and checking allowables at high levels of point design complexity is recommended to begin with.

NB: The generation of allowables for non-generic design features cannot be part of the material qualification process whose purpose has already been explained in the previous paragraph. In other words, qualifying a material does not imply the generation of all the data needed to size a given structure. It is very important to differentiate qualification testing for showing compliance with § 25.603 and structural substantiation testing for showing compliance with § 25.613. Merging both purposes may be misleading, with the risk of keeping qualification testing away from its original purposes.

§ 21.125 - Production Inspection System - Materials Review Board

The content of this regulatory paragraph is:

(a) Each manufacturer required to establish a production inspection system by § 21.123 (c) shall:

(1) establish a Materials Review Board (to include representatives from the inspection and engineering departments) and materials review procedures, and

(2) maintain complete records of Materials Review Board action for at least two years.

(b) The production inspection system required in § 21.123 (c) must provide a means for determining at least the following:

(1) Incoming materials and bought or subcontracted parts used in the finished product must be specified in the type design data or must be suitable equivalents.

(2) Incoming materials and bought or subcontracted parts must be properly identified if their physical or chemical properties cannot be readily and accurately determined.

(3) Materials subject to damage and deterioration must be suitably stored and adequately protected.

(4) *Processes affecting the quality and safety of the finished product must be accomplished in accordance with acceptable industry or United States specifications.*

(5) *Parts and components in process must be inspected for conformity with the type design at points in production where accurate determinations can be made.*

(6) *Current design drawings must be readily available to manufacturing and inspection personnel, and used when necessary.*

(7) *Design changes, including material substitutions, must be controlled and approved before being incorporated in the finished product.*

(8) *Rejected materials and parts must be segregated and identified in a manner that precludes installation in the finished product.*

(9) *Materials and parts that are withheld because of departures from design data or specifications, and that are to be considered for installation in the finished product, must be processed through the Materials Review Board. Those materials and parts determined by the board to be serviceable must be properly identified and re-inspected if rework or repair is necessary. Materials and parts rejected by the board must be marked and disposed of to ensure that they are not incorporated in the final product.*

(10) *Inspection records must be maintained, identified with the completed product where practicable and retained by the manufacturer for at least two years.*

This paragraph not only covers materials aspect, but also bought or subcontracted parts used in the finished product. As far as the materials issue alone is concerned, it provides the rule for ensuring that the proper materials and the proper processes will be used for the production of the part. Once the material is qualified, which means that, as required by § 25.603, compliance to a specification is shown, the difficulty with a new material is to define what has to be done in terms of incoming control in order to ensure detection of any deviation from the specified product. As already mentioned above, this is in the scope of the qualification exercise that key parameters of the material manufacturing process are identified and toleranced. It is also during this exercise that chemical, physical and mechanical properties able to detect deviations are to be identified and that a procurement specification can be established. While a qualification specification prescribes minimum performances to which various materials can comply, a procurement specification prescribes typical properties, with their associated tolerances, specific to each of these qualified materials. At the early stage of the development of a new material/technology, there may be poor knowledge about the key parameters actually governing performances. Then introducing the assessment of more engineering properties as part of the incoming control is highly recommended.

To be on one's guard against unexpected in-service troubles

When a new technology is 'type-certified' in an aircraft, all the actors, either belonging to the applicant or the airworthiness authorities, are convinced that the best has been done, from the current knowledge, to demonstrate that structural integrity will be achieved for the whole life of the aircraft. This means that all service life effects, such as long-term ageing for instance, have been correctly allowed for. The keynote lecture from Jens Hinrichsen, to which the author is responding, mentions two situations of completely unexpected in-service behaviours which have concerned Kevlar applications on fairings and bonded fuselage stringers, examples which are probably not the only ones. The same paper explains very well how the risk can be mitigated by preliminary explanatory developments plus starting to introduce novelties on less critical applications to get production and in-service experience. Despite all his or her efforts, nobody can claim to have investigated all long-term materials/technologies behaviours without the feedback of a significant in-service experience. Experience has shown that a fail-safe design with load path redundancies is essential to minimise the consequences of unexpected problems in service. No novel technology should be allowed to forget these principles.

Summary

Technological development prior to the programmes and involving all aspects such as reparability and maintainability, is essential to get new material applications to maturity. A step-by-step introduction of new technologies starting with less critical parts, is essential to reduce risks. How, over the last twenty years, Airbus has managed the introduction of composite materials in their programmes up to the level achieved today, deserves to be acknowledged. In the light of this experience, lessons have been learned that will benefit upcoming innovations, such as the introduction of Glare for instance.

When focussing on certification only, the following has to be pointed out regarding the introduction of a new material in a structural application:

- A building block approach, with abundant testing at the beginning, must always be adopted in order to get structure sizing fully under control, so that any deviation in the input variables has understandable effects and is manageable.
- The material/process qualification strategy has to be defined and implemented as early as possible in order to freeze the material in a definite configuration and to be able to control any deviation from its specification. This has to be done, even if there is only one material candidate to the qualification, and before structural substantiation testing for type certification starts.

- The endless race to the maximum structural performance and weight saving, which minimises all the margins, may prepare potential difficulties for addressing future issues for the life of the product.
- Doing one's best may not be sufficient to anticipate new materials/process in-service long-term behaviour. Fail-safe damage-tolerant design principles must be applied.

Response 3

Designing for risk: New materials and new approaches

Patrick T.W. Hudson

Department of Psychology
Faculty of Social and Behavioural Sciences
Leiden University

As a psychologist, I find it slightly daunting to be amongst so many engineers. I am more used to the dirtier end, i.e. maintenance shops at four in the morning, when all the things you do not want to know about happen.

What interested me when reading Hinrichsen's paper was the positive approach to risk and integrating design and risk. However, I would also like to sound a warning note since most engineers like to believe that they have found perfect solutions to their problems, when in fact they have usually created a new one instead.

The issue

Innovation is good and is a major driver of progress. However, if we look back, we find that most of the innovations are the results of attempts at solving problems that have been identified. One of the respondents today (ed.: September 26, 2001 – day 3 of the *Glare - The New Material for Aircraft* Conference – in the Aula Conference Centre of the Delft University of Technology in Delft, the Netherlands) commented that the problems just went away as a result of developments, which makes me feel slightly uncomfortable. It is clear that new materials like Glare bring significant benefits and solve many problems, but when you solve one problem, especially as the result of innovation, you do not necessarily solve all the problems. We have seen this on the flight deck, i.e. the change from the 'steam-driven' cockpit to the glass cockpit and now the second-generation glass cockpit shows that while things have definitely got a lot safer, they are not absolutely safe. We have often substituted old problems, i.e. ones that were solved, with new ones. Part of the problem is that there are humans in the loop, so that if we got rid of maintenance engineers and pilots then maybe flying would be a lot safer.

Three drivers

There are three main drivers. First, there is the demand for capital cost reduction in manufacturing and design, or in other words for the cheapest aircraft for the maximum load. Second, the operational costs must be kept under control, which is the airline's problem, but requires an aircraft that will last a long time, be cheap to run and be cheap to maintain. Third there is the requirement of a high level of safety. New technologies like Glare offer opportunities for all three of these.

Risk homeostasis

There is another problem, however, i.e. when people operate in terms of risk, whether it be as a committee, organisation or at a personal level, they do not actually think in terms of the real risk, since we do not really know what that is. We know what the historical risk was and we can guess what the future risk may be, but we have to learn that the numbers we produce are contaminated by the very fact that we know what they are. The real danger is that people act only on the risk that they perceive. Gerald de Wilde originally investigated this problem, which is known as risk homeostasis. It has been attacked for several reasons, but in the short term it proves to hold very true.

Therefore, while technological innovations improve the general safety of the system as a whole, risk homeostasis by the users (e.g. airlines, individuals, etc.) then takes advantage of the extra safety margin in order to take greater risks. Unfortunately, this means that once an engineer has built in all sorts of safety, someone else will do their best to undo it. It will not make it worse, in fact on balance things will usually get better, but we should be aware of this problem.

Airbus' philosophy

What I really remember from Hinrichsen's paper was that the Airbus philosophy shone through. It is quite clear that Airbus evolves its designs and that risk is always at the forefront, both in terms of the end user and the company itself. The problem is that the operators have been sold this concept of safety and may exhibit similar effects of risk homeostasis as Volvo owners who drive far too fast.

Design for maintenance

I was pleased to see a reference to maintainability. This is a relatively new idea since in the past a manufacturer would deliver a 'perfect', finished product and leave the maintenance up to the operator. This was a real problem. Investigation into aircraft crashes showed that engineering-induced crashes actually have a higher death toll than those caused by pilot error.

Figure 1: Artist's impression of the prototype of Airbus' future A380 [source: © Airbus Industrie].

In the last few years, the newer aircraft, e.g. Boeing's 777 and Airbus' A330, A340 and future A380, see Figure 1, have moved towards including the customers in the design process. Nevertheless, we have identified another problem, i.e. that the maintenance engineers can be left behind by the development of technologies.

Design for airworthiness

Building to airworthiness requirements by tailoring materials to specific tasks provides significantly greater resilience and usually at reduced costs. However, the problem is that there is the temptation to use new materials in other areas. It is a good thing that this process is controlled, since it could otherwise become dangerous. Unmanaged opportunism could lead to applications for which the material was not defined.

A salutary lesson

I would like to finish by considering what we could term the Airbus A100 or the Airbus A00, i.e. Concorde, see Figure 2. An interesting lesson to be learned from the development of Concorde is not to try to do too much at any one time. Evolve and develop step by step. The trick is to combine new materials and methods with conventional design and then later develop new design concepts using what have become conventional materials.

Concorde was an aircraft constructed within the manufacturing limits of its time instead of trying to develop new materials and new airframe concepts. The American SST programme on the other hand required too much from materials and concepts, as a result of which it never flew. Unfortunately, the political fall-out from this meant that Concorde never got to fly much either.

Conclusion

In conclusion, I would like to say that the talk from Airbus indicates a sensible approach. It is appropriate, risk-based and evolutionary. It is also useful to note the inclusion of maintainability concepts and a whole-system concept, which does not concentrate too much on operations or manufacturing, for example. However, when you do have new and improved developments, experience shows that people will always push the envelope in ways you never expected, and when they do that, I fear I may have to come and investigate.

Figure 2: Concorde was constructed within the manufacturing limits of its time [source: © British Airways].

Response 4

New technology and safety: Some moral considerations

Peter A. Kroes

Department of Philosophy
Faculty of Technology, Policy and Management
Delft University of Technology

(Note: The author wishes to thank the members of the Department of Philosophy of the Delft University of Technology for their comments on an earlier version of this paper)

Introduction

The events that took place on September 11, 2001 in New York and Washington give a new sense of urgency to the reflection on the role of technology in our society and in continuation thereof on the moral responsibility of engineers. Civilian planes were hijacked and used to destroy the World Trade Center and part of the Pentagon, killing thousands of innocent people. It is still much too early to assess all the possible consequences of these disasters. But clearly it was a black day for civilisation in general, and for civic aviation in particular. The safety of the system of passenger transport by air failed in a dramatic way. For a long time hijacking has been a threat to the safety of passengers and crews and all kinds of precautions have been taken to try to avoid it. But a new dimension has been added; hijacked aeroplanes have proven to be an effective means for terrorist attacks on high-rise buildings with casualties on an unprecedented scale.

The safety of air transport systems is a very complicated matter. They are complex socio-technical systems; they involve technological objects or 'hardware' (aeroplanes, airports, communications systems, etc), 'software' (flight procedures, air-traffic procedures, legal regulations, maintenance procedures, organisational procedures, etc.) and various kinds of actors (pilots, airline companies, maintenance staff, certification institutes, weather forecasters, insurance companies, financing institutes, etc.). All of these elements, each in their own way but also in the way they are tuned to each other, directly or indirectly influence the safety of the whole air transport system. In the case of the WTC disaster something went wrong, and questions are raised whether or not this accident could have been foreseen and possibly prevented by taking the appropriate precautionary safety measures.

It may be that this particular accident could have been prevented, but at the same time one may ask whether in general the risk of accidents involving modern technological systems can be avoided completely. Unfortunately, there is ample evidence to suggest that the answer is negative; the list of accidents is too long and too well known to be repeated here. And the list gets longer and longer; in less than two weeks after the WTC accident a chemical plant in a densely populated area in Toulouse exploded killing about thirty and wounding about a thousand people. It is indisputable that the use of (large-scale) technology generates new risks.[1] This observation raises various moral problems. What are acceptable levels of risk, and who is going to decide about these levels? What is a fair distribution of risks among people? And since modern technology plays such an important role in these accidents, the question arises what, if any, the moral responsibility of engineers with regard to (the prevention of) these risks is. In the following I will focus mainly on the last question. I will discuss the idea of Martin and Schinzinger that engineering projects may be interpreted as experiments on society, i.e. experiments on people, and that in analogy with medical experiments it is morally desirable to apply the principle of informed consent to these experiments. I start with some remarks about engineering ethics and then turn to the ideas of technological innovations as social experiments and of informed consent and their possible implications for the professional practice of engineers.

Ethics and engineering

It is a long-standing problem within ethics, understood as the philosophical discipline dealing with the study of morals, how to demarcate moral problems from other kinds of problems. Following Bernard Williams we will start from a conception of the moral that focuses on the relation between human beings, more particularly a conception of the moral that [2](p.12) '...relates to us and our actions the demands, needs, claims, desires, and, generally, the lives of other people...'. The more an action by one individual has consequences for the well-being of others, the more that action has moral value and the more it becomes important to ask whether that action is desirable, required, objectionable or forbidden, in short whether it is a morally good or bad action.

In line with this approach, the main question of engineering ethics may be phrased in the following way: 'How should engineers, in their capacity of engineers, act in case their actions have consequences for the well-being of other people?'[2] This formulation immediately raises the question of what it means to be an engineer. There are many different disciplines within engineering and within each discipline there are often different engineering profiles, e.g. research, design, management and consultancy. For our purposes it is not necessary to go into this problem. Various engineering disciplines have drawn up professional codes of ethics and almost all

[1] The German sociologist Ulrich Beck has characterised modern society as a risk society, see Beck [1].
[2] For more information about the field of engineering ethics see Martin and Schinzinger [3] and Harris, Pritchard and Rabins [4].

share a common moral obligation. In the code of ethics of the Accreditation Board for Engineering and Technology (ABET) this common element is phrased in the following way: 'Engineers shall hold paramount the safety, health and welfare of the public in the performance of their professional duties' [3] (p.342). In accordance with our interpretation of the notion of the moral, these codes of ethics stress the relation between the actions of engineers and other people, namely the public. Given this moral obligation to hold paramount the safety, health and welfare of the public, what lessons, if any, can be drawn from the WTC accident concerning the moral responsibility of engineers?

A technocratic response

The fact that the WTC disaster was caused by *deliberate* acts of terrorists should not lure us to the simple conclusion that in this case the question of moral responsibility is dealt with adequately and completely, with the observation that only the terrorists bear moral responsibility because of their morally objectionable intentions. Intentions of people are often considered to be important or relevant elements in moral issues, but they are not, it seems, the only ones. Suppose that indeed it had been the case, as a CNN news reporter initially announced, that the first plane had hit one of the towers of the WTC through a malfunctioning of the navigation system. Then the disaster would not have been caused by a morally objectionable intentional act. Surely that situation would not have meant the end of the discussion about moral responsibility – think of the Bijlmer disaster in the Netherlands in 1992. Many accidents involving modern technology do not involve objectionable intentions; the fact that they are unintended does not mean that there are no interesting moral issues to be raised.

Whether intended or not, when accidents take place, the moral discussion is usually driven by the question whether the accident could have been foreseen and by whom, and if so, whether it could have been prevented and by whom. In the case of complex socio-technical systems, such as civic air transport, these questions are very difficult to answer and as a result, the moral issues involved become very opaque. Could the WTC accident reasonably have been foreseen? Or, more generally, could accidents like the WTC disaster reasonably have been foreseen? On these matters opinions diverge, not in the last place because people disagree about the meaning of the notion 'reasonably'. Suppose it could reasonably have been foreseen, could the accident have been prevented? Some have suggested that these kinds of accidents are inherent to the kind of society we live in ('open society') and cannot be avoided. Others claim that the accident might have been prevented by a better functioning of the intelligence services, and again others claim it might have been prevented by simple technological measures, e.g. an improved cockpit door. As we remarked before, civic air transport is a complex system involving technological artefacts, human beings and social institutions, all of which contribute to the functioning and safety of the whole system. Consequently, measures to avoid accidents can be taken in the domain of technical artefacts (e.g. safer planes), the behaviour of human beings (e.g. better-trained staff) and social institutions (e.g. better regulations), or in the way these various elements are related to each other. So, many actors contribute to the safety of the system as a whole and when a safety failure occurs, it is often impossible to determine who is morally responsible for what. This is called the 'problem of the many hands'. When accidents involving modern technology happen unintentionally, a danger of the 'problem of the many hands' is that the whole issue of moral responsibility evaporates. Because it is not possible to trace the cause of the accident to a particular (combination of) 'act(s) of man', it becomes rather tempting to take a fatalistic attitude and consider it to be an 'act of God' or to be due to 'bad luck'. At that moment, moral analysis comes to a halt.

Among the many hands involved in an accident, we often find the hands of engineers. This is also the case in the WTC accident. Engineers are involved, for instance, in the safety of technological objects such as aeroplanes and high-rise buildings. But the safety of these objects is only one element relevant for the safety of the whole system of civic air transport. From a technological point of view, nothing specific appears to have gone wrong with regard to the safety of these technical objects. From an engineering point of view, therefore, a rather obvious response to the WTC accident is that the disaster was caused not by failures of technical subsystems, but by failures of social/human subsystems (failure of the security system for preventing hijacking or human failures in the sense of abuse of aeroplanes for unintended purposes). This diagnosis is often accompanied by a suggestion for a specific remedy: since the human factor is the most unreliable and error-prone element in the whole socio-technical system, try to avoid the occurrence of social/human failures by technological means. For instance, in the aftermath of the WTC accident it has been claimed that this kind of accident can be avoided with the help of advanced guidance and control systems, which would make it impossible to fly aeroplanes into buildings.[3]

This kind of response from the engineering community to the WTC accident will be called here the 'technocratic response'. From a moral point of view it places engineers in a rather comfortable position; it automatically portrays them as contributing to the prevention of morally bad effects or bad use of technology. But for several reasons this technocratic response is problematic. In the first place, it assumes that it is possible to make a clear-cut distinction between human and technological failures in the workings of socio-technical systems. This distinction is problematic. Take the example of the disastrous Challenger launch in 1986. On the one hand, it could be maintained that this accident was due to a technical failure, i.e. of an O-ring, on the other due to a human failure, i.e. a wrong decision making procedure at Thiokol. Moreover, the idea of a technical failure is itself problematic; many, if not all(?), technical failures are due to human failures, e.g. bad design by engineers, bad maintenance by technicians, etc. So it is not clear whether the notion

[3] This was claimed by Mulder from the Faculty of Aerospace Engineering of the Delft University of Technology, see *de Volkskrant*, September 13, 2001.

of a technical failure, as opposed to a human failure, makes sense at all.[4] In the second place, the introduction of new technology in order to eliminate human error may raise interesting moral problems. Under which circumstances, for instance, would a pilot be allowed to overrule an advanced guidance and control system? Here designers of these systems have to make decisions that may have far-reaching consequences for the safety of the people involved; these decisions are therefore not morally neutral. Finally, the technocratic response ignores the fact that with accidents like the WTC disaster, technology itself may be part of the problem in the sense that modern technology makes these accidents possible. The question has to be posed whether more technology will solve that problem. Is more technology not going to produce more risks? In the following section I will discuss an alternative to the technocratic response, which puts the issue of the moral responsibility in another, wider perspective.

Technological innovations as social experiments

In order to explore the moral responsibilities of engineers, Martin and Schinzinger proposed considering engineering as a form of social experimentation [3] (ch.3).[5] They believe that engineering is an inherently risky activity and shares some interesting similarities with standard experiments. In the first place, engineering projects, just as experiments, are carried out in partial ignorance. There are always uncertainties about the validity of models used in design, the characteristics of materials, etc. Secondly, there may be uncertainties about the end results of engineering projects. For instance, the effects of a new, decentralised water supply system on the social fabric of a village may be hard to predict. Finally, similar to experiments, good engineering requires the constant monitoring of the operation and effects of products, also in their context of use. This feedback is necessary to learn about the viability of engineering products, their adaptation to changing user requirements, their improvement, etc. What Martin and Schinzinger want to stress is that if engineering projects are considered as social experiments, then these experiments involve human beings, more in particular, they are experiments *on* human beings, and that in this way the focus of our moral analysis is put where it should be, namely on the people affected by technology [3] (p.67).

This is an interesting idea that may be applied not only to engineering projects, but also to technological innovations in general. The important point to note is that a technological innovation is not simply the introduction of some new technological artefact into an existing social structure that remains unchanged. The realisation of the function of that artefact requires a particular practice of (collective) human action. That practice may directly or indirectly change the existing social structure.

[4] Note that to appeal to technical failures as distinct from and opposed to human failures has far-reaching consequences for moral analysis: if an accident is considered to be due to a technical failure in this sense, then the issue of moral responsibility loses its meaning.

[5] The notion of social experiments used in the following is not to be confused with the notion of experiments in the social sciences.

In other words, technological innovations may induce social innovations (new forms of human behaviour, new institutions); new forms of socio-technical systems emerge. It is often very difficult to predict how these new socio-technical systems will develop and how they will influence the other social structures that are not directly related to the introduction of the new technology (higher-order/long-term effects). It is this aspect of uncertainty that turns technological innovations into social experiments.

There are also differences between these kind of social experiments and standard experiments, i.e. experiments conducted in scientific or technological laboratories. Martin and Schinzinger mention the absence of control groups in contrast to many experiments in medical and social sciences, for instance. Furthermore, experiments in science and technology are conducted to gain new knowledge whereas this is not the primary aim of engineering as social experimentation. They also point out that the principle of informed consent, which plays a paramount role in experiments involving human subjects in science and technology, is conspicuously absent in engineering as social experiments despite the fact that they also involve human subjects. Before we go into this point in more detail, let us dwell a little longer on differences between standard experiments and engineering as social experimentation. Experiments in science and technology are usually conducted under highly controlled circumstances. In order to prevent disturbing influences, experiments take place in closed environments (as much as possible), preferably within laboratory walls. The behaviour of these quasi-closed systems is controlled by varying relevant parameters. The situation with regard to engineering as social experiments is completely different. They are conducted within social systems that are commonly open to outside influences (disturbances). Because of this, there is much less control over the behaviour of these systems. Directly related to this is the fact that the behaviour of these open systems is much more difficult to predict. The outcome of engineering projects as social experiments may therefore be highly uncertain. Finally, the impact of failed experiments is quite different in nature for both types of experiments. If a standard experiment fails – because it is performed in a wrong way, or because the expected result does not occur – then this failure has only cognitive significance. But when an engineering project fails, this may have far-reaching impact on human lives.

As an illustration of the idea of engineering as social experimentation, let us have a closer look at the new material Glare and its application in the Airbus A380 aircraft. This new aircraft represents the next step in the evolution of ever-larger aircraft; it is intended to carry more than five hundred and fifty passengers. There is no experience with aircraft of this size and type. Its innovative features include an overall double-deck structure and the use of the composite material Glare for large parts of its fuselage. In the announcement of this conference, the Airbus A380 is described as a 'groundbreaking new aircraft'. From the paper by Jens Hinrichsen [5] it becomes clear that because of the use of advanced technologies and new structural designs, engineers are entering terra incognita here, although they do so very cautiously by taking small steps and building on earlier experience. Nevertheless,

Hinrichsen is clearly aware of the pioneering nature of this undertaking and of the risks that go with it. He remarks that 'a learning process has to be established for new technologies', that there are "teething troubles' experienced with almost every new technology', that 'there are 'broken bones' in the industry all around the world, resulting from applications of new technologies' and that it is necessary to 'mitigate risks from initial steps into new technologies'.

In his paper, Hinrichsen focuses on the safety of the aircraft itself and the uncertainties inherent in the application of new technologies. But let us draw the boundaries of the system under consideration somewhat wider and focus on the system of future civic air transport, of which this aircraft is going to be an element. What are going to be the consequences of the introduction of this new type of aircraft for civic air transport? Is it going to increase the scale of civic air transport? How will it affect the problem of environmental pollution? Is it going to affect the safety of air transport (will it be possible to perform all necessary security checks for so many people in a given time frame?)? What will be the consequences when a fatal accident with a fully loaded aircraft takes place? How will that affect the public perception of the safety of civic air transport and the behaviour of airlines, governmental institutions, the industry involved, etc? Clearly, the consequences of the introduction of this new type of aircraft are difficult to predict. So, on top of the uncertainties due to the application of new technologies, there are uncertainties about how the socio-technical system of civic air transport is going to absorb the introduction of this new type of aircraft. Because of these uncertainties this engineering project may be viewed as a social experiment.

Informed consent

If we assume that engineering projects like the Airbus A380, or technological innovations in general, are in a genuine sense social experiments, then these are experiments involving people. As Martin and Schinzinger observe, this opens up a new perspective on the moral and social responsibility of engineers. It brings the principle of informed consent into play. This moral principle has been developed within medical practice, originally to protect the interests of human subjects participating in medical experiments. It was intended to guarantee freedom of choice of test persons on the basis of sufficient information about the nature, setup and risks of medical experiments. Nowadays, the principle of informed consent is applied to any kind of medical treatment, whether experimental or not. The principle of informed consent is closely related to the idea of autonomy. Generally speaking, this principle states that humans have the right to live their lives the way they like, as long as they do not affect the well-being of others.[6] Within the situation of medical experiments (treatments) this principle is interpreted to mean that people have the

[6] This idea of autonomy goes back at least to Mill's famous treatise *On Liberty* that was originally published in 1860, see Gray and Smith [6] (in particular p.30-31). The above formulation stems from Zandvoort – who refers to it as the principle of self-determination –, see Zandvoort, *Self determination, strict liability, and ethical problems in engineering*, in: Kroes and Meijers [7] (p.220).

right to be informed about all relevant aspects and have the right to decide freely, on the basis of this information, whether or not to participate in an experiment or to undergo a medical treatment – assuming that this decision affects only their own well-being. Of course, the application of this principle in actual medical practice raises all kinds of problems (who has to provide the information? what is sufficient information? which information is relevant? what to do if the experiment is spoiled by informing the test persons about the nature of the experiment? etc.). These will not concern us here, however.

According to Martin and Schinzinger one of the differences between standard (medical) experiments and engineering projects as social experiments precisely concerns the principle of informed consent [3] (p.68):

> *But while current medical practice has increasingly tended to accept as fundamental the subject's moral and legal rights to give informed consent before participating in an experiment, contemporary engineering practice is only beginning to recognise those rights. We believe that the problem of informed consent, which is so vital to the concept of a properly conducted experiment involving human subjects, should be the keystone in the interaction between engineers and the public.*

If engineering projects are taken to be social experiments, then the analogy with medical practice strongly suggests that the moral obligation of engineers towards the public, as described in professional codes of ethics such as that of the ABET, requires the application of the principle of informed consent in engineering projects.

Is it indeed morally desirable or necessary to impose the principle of informed consent on engineering projects or technological innovations? An important difference between medical practice and engineering practice appears to put Martin and Schinzinger's position into question. In medical practice it is usually fairly obvious who will be exposed to the risks involved in the experiment or treatment and thus it poses no problem from whom informed consent should be taken. This is not the case for many engineering projects. Often it is very difficult, if not impossible, to predict whose lives will be affected when the project fails or what kind of negative side effects will occur, so it is not clear who to ask for informed consent. The only way to respect the autonomy principle in such situations appears to be to ask informed consent of all possibly affected, and in practice that often means all members of a given community. Apart from practical problems, this leads to the situation that everybody in that community may veto an engineering project by refusing his informed consent.

On these grounds it could be argued that for practical and/or principal reasons the principle of informed consent cannot be applied to engineering projects. If this line of argument stands up against criticism, and the principle of informed consent is rightly rejected for engineering practice, the problem remains how to arrive at a morally fair distribution of risks arising from engineering projects and technological innovations within society. If the principle of informed consent can not do the moral work in engineering practice that it is expected to do in medical practice – leaving aside all kinds of problems about how to implement this principle in medical practice –, then engineers have to face the question of which alternative principle should be adopted in order to arrive at a morally satisfactory solution of the distribution of risks generated by engineering projects.

Let us suppose that the analogy between medical and engineering practice holds; the experimental nature of engineering projects and respect for the autonomy of the people possibly involved requires application of the principle of informed consent. How to implement this principle in practice? This would not only require changes in engineering practice itself, but also in its wider context. Let me mention just two of those changes. In the first place, informed consent requires access to all relevant information. Secrecy is out of the question. This is not actual practice in most industries, especially not high-tech industries such as the aviation industry.[7] Much information about technical details, also information which is relevant for the safety of aircraft, is confidential. Access to all information is also important to assess the reliability of information. Informed consent does not imply that all information provided should be reliable, but in order to assess the reliability of information free access to all relevant information is crucial. In this respect it is interesting to note that the International Committee of Medical Journal Editors has recently decided to tighten the guidelines for authors [8]. In order to diminish the influence of the pharmaceutical industry on the publication of the results of research funded by that industry, authors now have to declare that they had full access to all relevant data of their research and that they take full responsibility for the reliability of the data and their analysis. This is to ensure scientific integrity, reliability of data and the safety of patients. Within medical practice, confidentiality of research data endangers the application of the informed consent principle. The same applies to engineering practice. The application of informed consent will require far-reaching changes within that practice because it severely constraints the possibility to keep information confidential.

But more than only changes within engineering practice would be necessary. The question arises whether decision procedures about engineering projects, within the private or public sector, obey the principle of informed consent. De facto this is not so: in many cases engineering projects are executed in spite of protests by interest groups. The principle of informed consent puts a very severe constraint on the public and private decision procedures with regard to engineering projects. The informed consent of all people possibly affected by those projects is necessary. But take, for instance, the risks involved with the introduction of new, large-scale public infrastructures. How do we decide who is potentially exposed to its risks? It would be a mistake to assume that this consent is given implicitly in case the decision to introduce the new infrastructure is taken in a democratic way. Most democratic

[7] For a discussion of the right to be informed and the role of the law in informing the public about risks related to technology, see the exchange of ideas between Zandvoort and Vlot in Kroes and Meijers [7] (part entitled *A dialogue on engineering design and law*, p.193-250).

collective decision procedures are based on some form of majority rule, and therefore do not respect the principle of informed consent. This principle implies that decisions with regard to engineering projects can only be made on the basis of unanimous consent of all involved. That could easily bring technological development to a halt. Thus, if strict adherence to the principle of informed consent is a necessary condition for the fair distribution of risks associated with new technologies, then there is a real danger that it does so in a trivial way, namely by not allowing the introduction of these new technologies and their risks. That itself raises interesting moral problems, for possible benefits of new technologies are not realised. The application of the principle of informed consent for engineering projects has, from a moral point of view, its own drawbacks. It is not obvious that it offers a morally acceptable solution to the problem of the fair distribution of risks due to technology.

If engineering projects are considered to be social experiments, and there are good reasons to do so, then engineers are some of those conducting experiments on people. This brings with it moral responsibilities toward those people. Within medical practice the principle of informed consent is generally taken to be necessary for conducting experiments in a morally acceptable way. It remains an open question whether, and if so in what form, the principle of informed consent can do the same moral work in engineering practice.

References

[1] U. Beck, *Risk Society; towards a new modernity*, London: Sage Publ., 1992.

[2] B. Williams, *Ethics and the limits of philosophy*, London: Fontana Press, 1993.

[3] M.W. Martin and R. Schinzinger, *Ethics in engineering*, New York: McGraw-Hill, 1989.

[4] C.E. Harris, M.S. Pritchard and M.J. Rabins, *Engineering ethics: concepts and cases*, Belmont: Wadsworth Publ. Co., 1995.

[5] J. Hinrichsen, *The material down-selection process for A3XX*, preprint, 2001.

[6] J. Gray and G.W. Smith (eds.), *J.S. Mill On Liberty in Focus*, London: Routledge, 1991.

[7] P.A. Kroes and A.W.M. Meijers, 'The empirical turn in the philosophy of technology', in: *Research in Philosophy and Technology*, Volume 20, Amsterdam: JAI, 2000.

[8] *The New England Journal of Medicine*, Vol. 345, No. 11: p.825-827.

Emeritus Lecture Professor Vogelesang

The integration of academic education and research and development

Emeritus lecture
held on September 26, 2001
at the Delft University of Technology

by

Prof.ir. L.B. Vogelesang

'The scientist explores what is,
the engineer creates what had not been.'

Theodore von Karman

Emeritus lecture

The integration of academic education and research and development

L.B. (Boud) Vogelesang

Faculty of Aerospace Engineering
Delft University of Technology

Mijnheer de Rector Magnificus en overige leden van het College van Bestuur,

Geachte collegae hoogleraren, docenten en medewerkers van de Universiteit,

Geachte dames en heren studenten,

Beste familieleden, vrienden en collegae van buiten de Universiteit,

Dear friends and colleagues from abroad,

Valued listeners,

Today I have the honour of addressing you on the occasion of my retirement from a period of education and research in the field of aerospace engineering at the Delft University of Technology.

The Delft University of Technology

During my inaugural speech on June 9, 1993, I attempted to present a vision of the role of the technical university within our society. The society to which I referred was still somewhat limited. At the time, we focussed mainly on the Dutch society, feeling that we were largely responsible for the future of Dutch industry. Much has changed since then. Enormous advances in information technology have led to a globalisation of our societies and the co-operation of universities around the world.

The Delft University of Technology wants to belong to the world's top-ranking research universities in the field of science and technology. I support this ambition wholeheartedly: every self-respecting university must, of course, contribute to the efforts to push the bounds of science by fundamental research.

On the other hand, restricting ourselves to carrying out only fundamental, science-based research, would be to deny the true nature of research across the whole field of science and technology. This is, however, the current trend within what are known the world's top technical universities. I am of the opinion that this is not a route the Delft University of Technology should take. Delft is known the world over for the technical products its engineers produce, such as the Delta works, the construction of harbours, the design and manufacture of ships and aircraft, electrotechnical products, exploration for oil and gas, process technologies, etc. These advanced products are the basis for the respectable reputation the Delft University of Technology has earned throughout the world.

Prometheus.

Research always requires a goal. We want to know what we don't know, do what we can't do, create what doesn't exist, and all with our society in mind. Between 1605 and 1608 Simon Stevin van Brugghe published a series of teachings he gave Prince Maurits from 1593 onwards, entitled *Wisconstighe Ghedachtenissen* (trans.: Mathematical Notions). One of his notions concerned the 'mixing of reflection and deed', the combination of theoretical contemplation and practical execution. According to Stevin it is impossible to practise a craft without an understanding of the theory behind it. As the motto of the Dutch Royal Institute of Engineers (KIVI) says:

'Scheppend denken, denkend doen'
(trans.: 'Think creatively, create thoughtfully').

Education and research are inseparably joined. There can be no education without research, nor research without education. This is my premise. The Delft University of Technology is a technical university; 'university' stands for scientific education and science-oriented research, and 'technical' stands for object-oriented research (design).

This trinity forms the foundation of a technical university. We therefore need to obtain a good balance between science-oriented research, object-oriented research and scientific education.

The role of the Delft University of Technology is no longer in question. This role is prominently international. But let us not allow ourselves to be trend followers, instead we should be trendsetters. Scientific publication should not be the only thing that counts, pioneering design is equally important.

To assure a leading position in the European educational market means competing with the best European universities. This requires a clear education and research strategy, transparent leadership and output-controlled process organisation.

Faculty of Aerospace Engineering

The Faculty of Aerospace Engineering supports the university-wide mission wholeheartedly. It goes without saying that a self-respecting university must contribute to the advancement of science through important achievements in fundamental research. The faculty owes its respected international position to the merging and integration of various fields of study; this process is essential to integral aircraft and spacecraft design. The faculty's unique character stems from its object-oriented nature, in other words, its primary focus on air- and spacecraft throughout a comprehensive set of relevant in-house fields of study. Insight and overview, interconnectivity and collaboration are all pivotal notions for both faculty staff and aspiring students within the faculty. An essential aspect of being a scientist in a scientific engineering environment is that progress is made as a result of intense interaction between theory and practice.

There is no doubt about the striking success of the faculty, mainly due to its continuous striving for a balance of responsibilities – scientific research, object-oriented research and system integration in addition to academic education.

Faculty of Aerospace Engineering.

Chair Aerospace Structures and Materials: A typical Delft Chair

'Imagination is more important than knowledge.
Knowledge is limited. Imagination encircles the world.'

Albert Einstein

In my chair, besides research, the education of young creative MSc- and PhD-students is our main incentive, the red line through our large Structures and Materials Laboratory'

Education by performing research. Advanced research keeps our lectures up to date and highly stimulates our student design projects, which are performed in every year of our curriculum.

Structures and Materials Laboratory.

Static and dynamic test machines.

An engineer as we see him/her is a creator of technological objects using knowledge and methods derived from science. Our title *ingenieur* – according to our USA ABET assessment equivalent to the USA MSc – is derived from ingenious: an ingenious, creative person. Our goal is the training of 'top-level' *ingenieurs* with a broad fundamental knowledge base and a market-oriented approach: a materials engineer who knows about the design and application of material systems in relation to structural design and fabrication.

To educate this kind of engineer we created a self-supporting Structures and Materials Laboratory and invited the industry to work with our students and staff in our laboratory on pioneering research projects.

No pure scientists who are not able to co-operate with other disciplines and are not able to set results in perspective.

'...and we can save 900 Lira by not taking soil tests.'

The duty to synthesis.

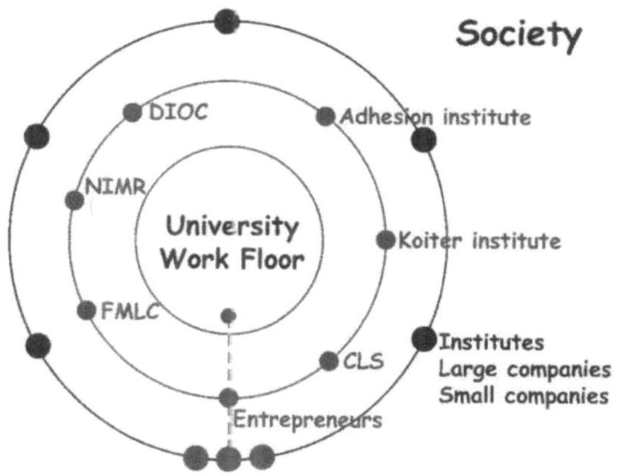

The university work floor network.

No computer screen engineers, but engineers educated in a realistic environment; we call it:

the 'University Work Floor'.

So we prefer to let our students perform their MSc- and PhD-research in our own laboratory rather than sending them to work in the industry.

Our work floor is a technology and science shop-floor for the industry. This is an advantage for both parties; the industry gets access to a source of technological knowledge and creativity, the university can keep its laboratories up to date, while the students work on realistic projects. It is a co-operation of equal partners. A leading university never needs to be afraid of loosing its academic freedom.

Our education always stays the red line through our science- and object-oriented research.

The university work floor, including the Structures and Materials Laboratory, stands at the centre of a network, with lively contact between the university and the community around it.

At the centre lies the university work floor itself, comprised of the Structures and Materials Laboratory, graduate students (MSc), post-graduate students (PhD) and scientific and technical staff. Surrounding this are a number of self-supporting institutes that function as windows to the outside world. These institutes are responsible for the transfer of knowledge and acquisition of new projects. At the moment these are:

- Foundation Fibre Metal Laminates Centre of Competence (FMLC), an equal co-operation between the National Aerospace Laboratory (NLR), Stork Fokker Aerostructures (Fae) and Delft

- Centre of Lightweight Structures (CLC), a co-operation between the Netherlands Organisation for Applied Scientific Research (TNO) and Delft
- Adhesion Institute Delft, an interfaculty institute
- Koiter Institute of Computational Mechanics, an interfaculty institute

The outer circle contains our permanent research partners, national and international companies, research institutes and governmental organisations. Crossing the circles from the inside out is a stream of young entrepreneurs. These entrepreneurs remain attached to the 'umbilical cord' of the 'womb' (the work floor) for as long as necessary until becoming companies in their own right in the outer circle.

The chair follows a top-down approach as regards most of the research.

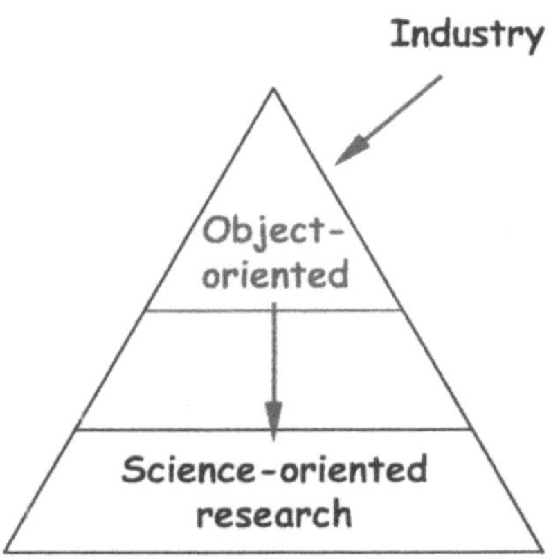

Top-down approach.

Society (industry) has a need for design capabilities or new developments that need to be undertaken. These requests are usually made via the institutes and end up reaching the university work floor. Projects that can be solved using ready-made solutions and knowledge, those which have already been developed, are usually undertaken by the institutes, where many contract researchers (*toegevoegd onderzoekers*) from the university will work on them. Projects that require the development of new knowledge and new techniques are usually passed on to the scientific university staff and post-graduate students. Graduate projects are mostly performed in both groups. This leads to the education of the *ingenieur* with a broad fundamental background as well as a market-based attitude.

I have heard critics comment that this system looks very similar to a higher technical education (*HBO*). This is not the case, however. Both *HBO* and the technical university educate engineers, both design products, but the *HBO* engineer will use existing tools to do this, whereas the TU graduate should also be able to use his science-oriented background to invent new and innovative tools to help create new solutions and pioneering designs.

A vertical projection of the preceding picture makes this clear. The object-oriented research is fully embedded in an environment of science-oriented research.

MSc-projects are completed in one year, while PhD-projects take four. To keep the quality of these projects as high as possible, the students need to have a modern and fully equipped laboratory at their disposal. This is essential for efficient research.

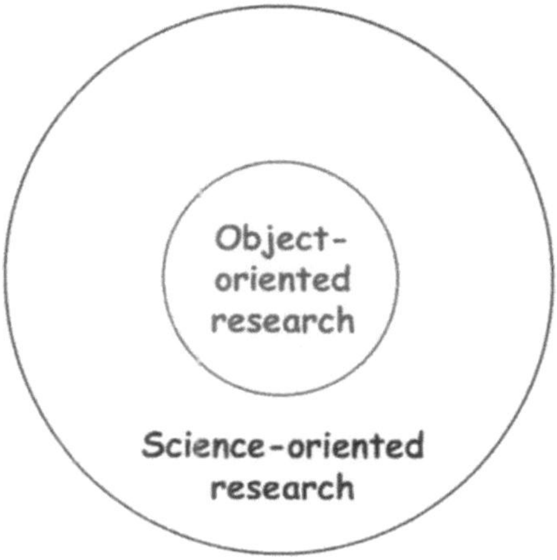

Object-oriented research embedded in science-oriented research.

The chair has a large national and international network that goes beyond aerospace engineering. Lightweight constructions, the expertise of the chair *pur sang* are becoming important in many other industries and have a significant potential throughout the whole transport and civil engineering sectors. This specialisation derives its specific characteristics from the stringent demands made by the aerospace industry. The product weight must be as low as possible, while carrying as large a load as possible and using as little fuel as possible. At the same time, the structure must be extremely reliable and require only efficient and cheap servicing and repair. The need for long life spans means that modern aircraft must be durable, in other words free from cracks and corrosion problems. They must also be resistant to damage and be damage-tolerant; the structure needs to have a fail-safe character. Even under rare and extreme circumstances, the aircraft should not fail.

This opposition, high safety versus lightweight construction, has led to the development of highly specialised materials within the aerospace industry, and thereby exceptional construction concepts and manufacturing techniques. In particular one can name:

Materials:

- hybrids such as fibre metal laminates
- fibre-reinforced plastics
- new aluminium alloys

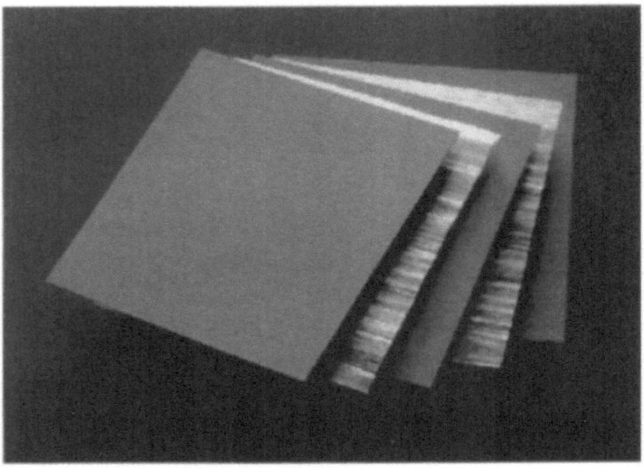

Glare build-up.

Structural concepts:

- thin-skin, self-supporting shell structures
- sandwich structures
- space frame structures

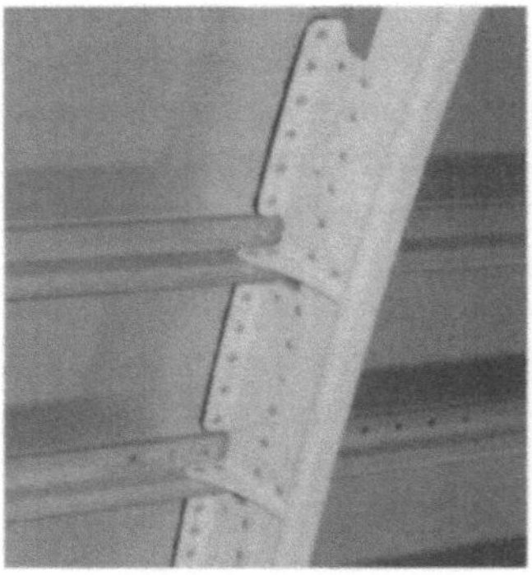

Thin-skin structure.

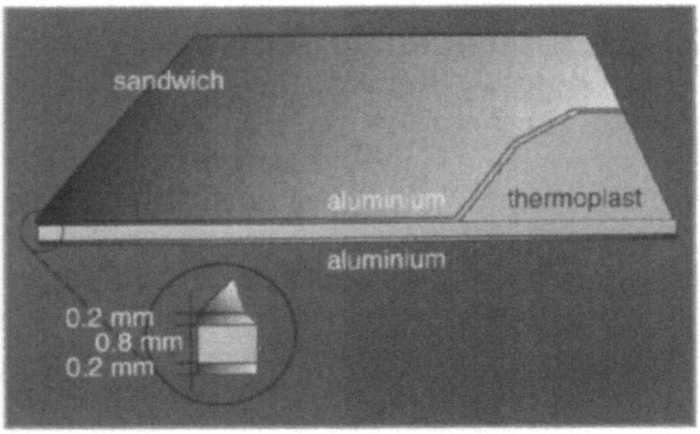

Sandwich structure.

Manufacturing techniques:

- splicing concept for hybrids
- filament winding of large components
- vacuum injection moulding
- advanced forming techniques
- advanced joining techniques

Splicing concept

Filament winding.

Adhesive bonding.

The fascination of technology

'In fact we have to give up taking things for granted, even the apparently simple things. We have to learn to understand nature and not merely to observe it and endure what it imposes on us. Stupidity, from being an amiable individual defect, has become a social vice.'

J.D. Bernal, *New Scientist* (taken from *The New Science of Strong Materials*, J.E. Gordon)

Technology has been a formidable force within society during the last two centuries, and this is still the case. Technological innovation has greatly altered our way of life. Such innovation is directly linked to economic growth, with a growing level of supply and demand and longer life spans in the western world. Well-being and prosperity are largely the result of the innovativeness of a country's business community. A lack of regenerative business leads to poverty and unemployment. Only the presence of regenerative business allows us to determine our own paths, in other words to retain our personal freedom and responsibility (KIVI workgroup 'Regenerative Business'). The universities in particular have a special role to play in this. They are the breeding place for the young, creative engineers who will form the backbone of the regenerative industries as well as being the breeding place for new developments through long-term strategic research.

President Marvin Goldberger of Caltech University says:

'Select the very best people, give them the very best facilities and stand aside.'

This is how the Structures and Materials Laboratory was created in Delft.

Smart wing.

The laboratory spends most of its time developing 'products', often up until the prototype stage. Such products include aircraft and spacecraft components, robot arms, predictive computational models (fatigue, residual strength), new material systems, new production techniques and new design tools (CAD, FEM, etc.).

We concentrate on advanced products with a high added value and acceptable labour costs. This inevitably leads to the development and application of new materials and clever design and production techniques.

One should not concentrate on only one group of materials when designing products. Lightweight structures are usually built up from various materials and use a variety of different joining techniques.

Integrated and modern design and production methods make an important contribution, allowing us to compete in an environment in which the competition is becoming more global. Applying new techniques helps us to overcome the handicap of the traditionally high Dutch wages and our social, economic and environmental constraints.

Most of the research in our chair is performed in the Structures and Materials Laboratory. The research efforts of the laboratory have three cornerstones:

- Science-oriented research: Successful application of new materials and design strategies can only be achieved if based on a thorough scientific understanding of the mechanical, physical and chemical aspects of materials and the optimal layout of structures.
- Integration of various disciplines: The laboratory has the knowledge, skills and equipment to cover the complete development of a structure: from materials science, structural design and manufacturing techniques to the fabrication and testing of full-scale components.
- Close co-operation with industry: The laboratory has a strong design-oriented approach. Input and questions from the industry are essential to guide the research, which is directed towards the gathering of engineering knowledge for the solution of practical problems.

The expertise of the laboratory covers an area from micro-mechanics of materials via design and manufacturing techniques up to full-scale testing of components. A thorough knowledge of and insight into the relationship between micro-structure and macro-properties of materials is of increasing importance when optimising the application of materials in constructions. This relationship is pursued experimentally, in combination with model development. The material behaviour that has been investigated includes the resistance against mechanical loading, both static and dynamic, durability, workshop properties, forming and environmental consequences like recycling.

The tendency towards more advanced materials, more powerful computational tools, modern design methods, more flexible and computerised manufacturing techniques and, last but not least, destructive and non-destructive inspection, requires an integration of the various disciplines involved.

There is now a strong interrelationship between material selection and properties, structural design and processing.

A representative selection of some of our research topics is given in the following.

Development of fibre metal laminates

The development of this family of materials is a typical example of a successful co-operation between students, staff and industry on the university work floor. This project demonstrates the strength of our philosophy based on education by performing research. A new material for the aerospace industry developed by students in a university laboratory in which object-oriented research (design) is supported by science-oriented research.

The FML-programme concentrates on the development of hybrid laminates for structural applications. Research is aimed at a successful adoption of the material by the aircraft industry. For that reason the group works together with FMLC, NLR, Fokker Aerospace and Airbus Industrie.

After 20 years of intensive research the real breakthrough came last year when Airbus Industrie chose the Glare variant of FML for a large part of the fuselage structure of their new A380 high-capacity aircraft.

A380 high-capacity aircraft.

Through Glare, the university work floor reached international fame and realised a 'technology-mature' hybrid structural material with a combination of excellent material properties.

Furthermore, the splicing concept offers the possibility to increase the size of FML structures without decreasing the excellent residual strength and fatigue properties. Integration of aircraft production steps in the production of FML in combination with the application of the splicing concept yields a cheaper aircraft – in terms of production, operating and maintenance cost – at an increased safety level: damage tolerance is built into the material and Glare also has a high burn-through resistance.

Large fuselage panel without any joints.

A310 MRT in-flight Glare test.

Damage-tolerant repair techniques for pressurised aircraft fuselages

The need for good repair techniques predates powered flight and continues to be an integral part of flying today. Structural repairs on commercial airliners and military transports are most typically required for fatigue cracking, corrosion and incidental damage such as impact. For military aircraft, battle damage joins the above list.

In 1991 an ongoing joint project with the USAF was initiated to develop a computer design tool – with the acronym 'CalcuRep' –, which enables a user to accurately design a safe bonded repair in which all affecting variables are accounted for.

An inherent fatigue-, corrosion- and impact-resistant material, like Glare, may help to ensure a damage-tolerant solution for bonded repairs.

Compared to mechanical fastening, adhesive bonding provides a more uniform and efficient load transfer into the repair patch and can reduce the risk of high stress concentrations caused by riveted repairs.

The effects of different temperatures and moisture levels on the bonded repair efficiency were investigated. This meant gaining knowledge of the effects of different environments on each component of the repair through extensive testing, even in flight, backed up by finite element modelling. Techniques like Ultrasonic Scanning and Scanning Electron Microscopy (SEM) were used to visualise and qualify those environmental influences.

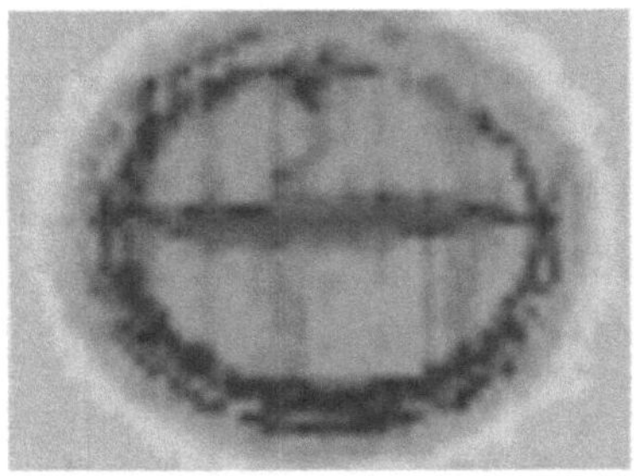

Delamination around crack under bonded repair.

Bonded Glare patch repair on C-5 Galaxy.

Repair of the Panorama Mesdag

The water damage inflicted on the huge and world-famous Panorama Mesdag, a true piece of Dutch national heritage, by a heavy thunderstorm on June 2, 1983, clearly meant that something had to be done.

The Panorama Mesdag is one of the few 19th-century panoramas that still exist; the others in Europe were given up after such damage. In their search for knowledge, the restorers called on our laboratory with a view to applying advanced methods of repair to the Panorama as used for aircraft.

Panorama Mesdag.

The backing chosen was a nylon fabric. We carried out computer calculations to estimate the stresses that would occur and the shape of the canvas before and after lining it with the nylon.

The Panorama canvas has a unique concave 'hourglass' shape, which was not to be affected by the restoration.

Calculations showed that the peculiar shape would be preserved. Advanced repair techniques were developed using heat blankets and a vacuum frame to apply the accurate uniform pressure during the bonding process.

To master the lining process, a model of a Panorama segment having a height of 9 metres and in the same double-curved shape was built at the laboratory.

An extensive durability test programme ensures that the repair will hold for at least 50 years. The scene of the peaceful beach of Scheveningen of 1880 has thus been preserved. An official opening by Queen Beatrix marked the completion of the successful high-tech restoration project.

Full-scale test setup.

Development of the DART, Delft Aerospace Re-entry Test demonstrator

The DART is a small experimental space capsule, designed to investigate the extremely hot gas flow around the capsule during re-entry into the atmosphere.
The capsule will be made from the super-alloy PM 1000. This is an alloy based on nickel and chromium and it can resist temperatures up to 1200 °C. When it oxidises, a thin oxide layer will be formed that will protect the structure against further attack.
To avoid overheating the metal, an ingenious water protection system has been developed. During re-entry the water vaporises, the steam is discharged, and so overheating will be avoided. Calculations have shown that only 10 litres of water is enough to discharge the heat of the free fall into our atmosphere.

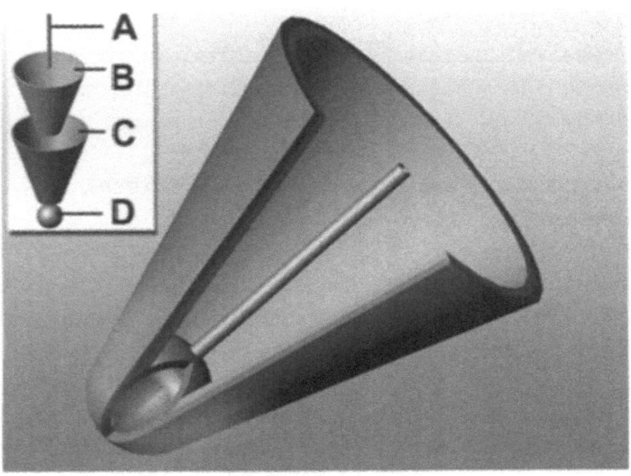

DART re-entry capsule.

The development of an ultra-light sustainable concept car

The Dutch-EVO – EVO stands for evolution – prototype study of a car, has been undertaken to stimulate innovative, multidisciplinary object-oriented research. The parties who initiated the programme are the Faculty of Industrial Design (design and control), Applied Earth Sciences (product life cycles) and the Faculty of Aerospace Engineering (structural design, aerodynamics and safety) in co-operation with TNO.
The framework for the Dutch-EVO consists of: minimal fuel consumption (2.5 litres/100 km), 4 passengers and luggage, lightweight design (mass 400 kg), environmentally friendly and application of renewable material resources. This goal cannot be achieved with the available technology. New techniques have to be developed, a special aerodynamic shape, advanced designing techniques, especially for impact, new materials and new suspension techniques. The project aims for the realisation of a full-scale operating prototype.

Dutch-EVO.

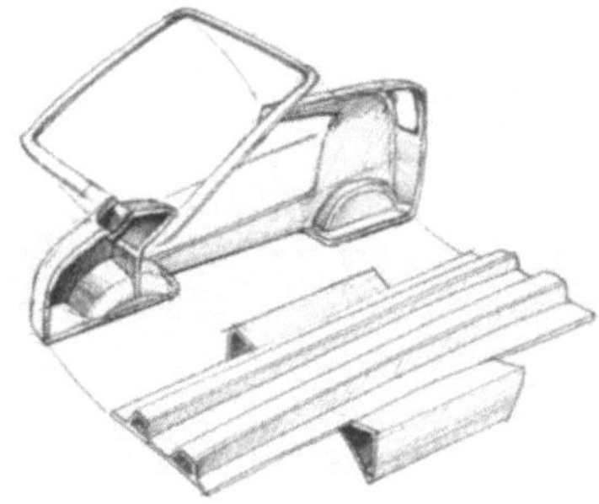

New structural concepts.

Teamwork

Teamwork is the ability to work together towards a common vision. The ability to direct individual accomplishment towards organisational objectives.

It is the fuel that allows common people to obtain uncommon results.

It is not my style to look back; I am much more interested in the future. I strongly believe in a prosperous future for my own group: students, staff and institute members. What a fantastic team! Without our team spirit and a strong mutual belief in a risky research project, Glare would never have become a success.

My grandfather advised me, when he heard about my choice for a scientific carrier, never to take the common road. He was a wise man. So I chose, remembering his advice, the road of hybrid materials. And that was not an easy one.

Part of the research group of Structures and Materials Laboratory.

Belief in the future is more important than predicting it. An engineer must never let himself attempt to predict the future. That is impossible, since almost everything is possible.

Flying by manpower is impossible according to Giovanni Alfonso Borelli in his famous book *Matu Animation* (trans.: About the Motion of Animals). He was wrong and I always taught my students to be stimulated by such absolute statements to prove the opposite. Daedalus tried to do so. The myth of the Icarus story is intriguing and I used to ask my first-year students about the reason for the accident. Was it pilot error or the wrong structural design?

Maybe Daedalus and his son forgot to team up with others. Then they might have come to the conclusion that the available materials and joining techniques were not good enough and more research had to be done first.

Anyhow, with Glare we did not make that mistake but teamed up with specialists, the best in their field: AKZO Nobel, 3M, ALCOA, Fokker, Airbus Industrie, NLR and, with flying colours, the Structural Laminates Company (SLC).

'Icarus Flying' by Themis Tsiromis.

Gossamer Kondor, manpowered flight.

Closing remarks

I have now come to the end of my speech and would like to thank my foreign guests for their patience as I switch to my mother tongue to say my final words of thanks.

Dames en heren,

Het is heel lang geleden, namelijk in september 1957, dat ik mij meldde als 1^e-jaars student aan de toen nog Technische Hogeschool Delft. Een rekensommetje leert mij dat ik bijna driekwart van mijn huidige leven verbonden ben geweest aan de TU Delft. Ik mag dus met recht zeggen dat ik een echte Delftenaar ben, en dat voelt heel erg goed.

Oude Kerk *Delft*.

Bij mij is tegenzin in het werk nooit echt aan de orde geweest. Natuurlijk had ik ook mijn mindere momenten, maar dan was daar altijd de koffietafel in het laboratorium, waar mijn studenten zaten, jonge veelbelovende mensen, altijd enthousiast met een vast vertrouwen en zin in de toekomst. Dan sloeg mijn slechte stemming snel over en besef je waar het echt om gaat.

Dames en heren studenten, het was voor mij een voorrecht om voor u te mogen werken. Ik zal u missen.

In mijn intreerede heb ik de mythe van Sisyphus aangehaald. Sisyphus als symbool voor de moderne ingenieur. Voor de mensheid zal nooit een moment komen om op de top van de berg te rusten. Ons werk zal nooit zijn voleindigd. Dat stelt mij als emeritus hoogleraar weer een beetje gerust. Ik blijf dan toch maar in de buurt, zij het op bescheiden afstand.

Voor mij was natuurlijk de meest vertrouwde omgeving die van de leerstoel en van het laboratorium. Creatief en vooral grensverleggend bezig zijn, ik heb ervan genoten. En dat in een faculteit die tegen de verdrukking in – denk maar aan het faillissement van Fokker – gestaag blijft groeien. Dat komt vooral doordat wij voor onze eigen weg kiezen, en niet meelopen in de grote massa. De faculteit verkiest trendsetter te zijn en niet trendvolger.

Aan veel mensen ben ik dank verschuldigd. Zowel binnen als buiten de TU-gemeenschap. Gaarne zou ik al die namen willen noemen waarmee ik heb samengewerkt en die zo enorm hebben bijgedragen aan mijn werkgeluk. Helaas is dat nu niet mogelijk. Een uitzondering wil ik maken voor Jaap Schijve, mijn leermeester en nog steeds betrokken bij de leerstoel, voor Jan Willem Gunnink, directeur van SLC, nu directeur van FMLC, mijn directe partner en motor bij de ontwikkeling van vezel/metaal laminaten, voor Theo de Jong, die als decaan ons afschermde van de bestuurlijke perikelen, en zoveel heeft bijgedragen aan het goed functioneren van de faculteit, voor Ad Vlot, mijn opvolger met zijn enorme inzet, kunde en loyaliteit, voor René de Borst en Adriaan Beukers, mijn inspirerende collega's, and last but not least for Jens Hinrichsen, director Structural Engineering, Airbus Large Aircraft Division, promoter of Glare within Airbus Industrie, who highly stimulated my research team. During the opening of the Fibre Metal Laminates Centre of Competence on May 6, 2001, Jens gave a presentation entitled *Glare, how to get an idea flying*, and I like to show you now two of his last slides.

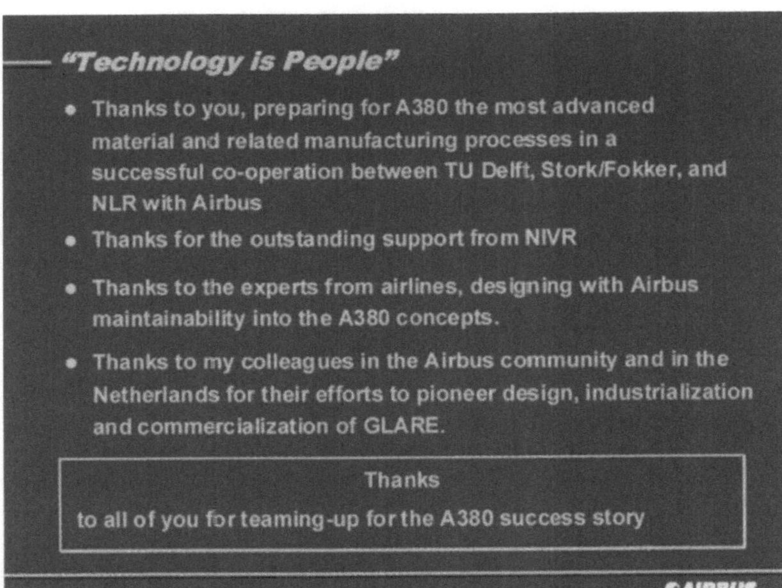

Teaming-up with Airbus, NLR and Stork Fokker.

Teaming-up with Airbus, NLR and Stork Fokker (cont.).

Al bijna 40 jaar deel ik mijn leven, en ook zal ik mijn toekomst delen met Vonnie. Dankbaar ben ik mijn vrouw voor haar aanmoedigingen, haar opofferingen en haar begrip. Zonder haar had ik dit werkstuk nooit geklaard. Het is dan ook mede haar werk.

Ladies and Gentlemen, I thank you for your attention.

Ik heb gezegd.

Sponsors

Main sponsors

- Delft University of Technology
- Airbus
- Fokker Aerostructures B.V.
- Aviation Equipment, Inc.
- FMLC - Fibre Metal Laminates Centre of Competence
- European Office of Aerospace Research and Development, Air Force Office of Scientific Research, United States Air Force Research Laboratory

Co-sponsors

- Ministry of Economic Affairs
- KLM Royal Dutch Airlines
- Advanced Glassfiber Yarns LLC
- SP aerospace & vehicle systems
- Embraer - Empresa Brasileira de Aeronáutica S.A.
- City of Delft
- Advanced Lightweight Engineering BV
- National Aerospace Laboratory NLR
- Netherlands Agency for Aerospace Programmes NIVR
- 3M Aerospace and Aircraft Maintenance Department
- Kluwer Academic Publishers
- TNO
- ASCO Industries NV
- Pechiney Rhenalu
- NAG - Netherlands Aerospace Group
- SAMPE

STORK®
Fokker
Glare®
A380
A380
Vision becomes reality
Fokker Aerostructures is the design and manufacturing partner to Airbus for a.o. Glare fuselage panels
Fokker Aerostructures B.V. P.O. Box 1 3350 AA Papendrecht The Netherlands
Tel. +31 78 64 19 399 Fax. +31 78 64 19 799 http://www.fokker.com
Fokker Aerostructures B.V. is a member of stork aerospace

FMLC is the Centre of Excellence, which supports customers world-wide, to develop and manufacture Fibre Metal Laminates applications.
A380
FMLC
Kluyverweg 6
2629 HT Delft
The Netherlands
tel. +31 15 251 47 40
fax +31 15 257 07 86
e-mail info@fmlc.nl
FMLC
Fibre Metal Laminates
Centre of Competence

MIX
Papier aus verantwortungsvollen Quellen
Paper from responsible sources
FSC® C105338

If you have any concerns about our products,
you can contact us on
ProductSafety@springernature.com

In case Publisher is established outside the EU,
the EU authorized representative is:
Springer Nature Customer Service Center GmbH
Europaplatz 3, 69115 Heidelberg, Germany

Printed by Libri Plureos GmbH
in Hamburg, Germany